The Great Telecom Meltdown

For a listing of recent titles in the *Artech House Telecommunications Library*, turn to the back of this book.

The Great Telecom Meltdown

Fred R. Goldstein

ARTECH
HOUSE

BOSTON | LONDON
artechhouse.com

Library of Congress Cataloging-in-Publication Data
A catalog record for this book is available from the U.S. Library of Congress.

British Library Cataloguing in Publication Data
Goldstein, Fred R.
   The great telecom meltdown.—(Artech House telecommunications Library)
   1. Telecommunication—History   2. Telecommunciation—Technological innovations—
   History 3. Telecommunication—Finance—History
   I. Title
   384'.09
   ISBN 1-58053-939-4

Cover design by Leslie Genser

© 2005 ARTECH HOUSE, INC.
685 Canton Street
Norwood, MA 02062

International Standard Book Number: 1-58053-939-4

10 9 8 7 6 5 4 3 2 1

Contents

Preface

The economic boom of the late 1990s included huge investments in the tele-communications industry and related sectors. It was followed by a downturn of unusual severity, which reduced total paper wealth by trillions of dollars, cost many thousands of jobs, and saw some of the biggest bankruptcies in history. While there certainly was a general business cycle at work, the downturn in tele-com was not just a cyclical correction, itself a healthy event that routinely shakes out the weakest players. The downturn was unusually severe, impacting many well-established as well as young companies; the price pressures that resulted from so many distressed vendors then put an impossible squeeze on the profit margins of many of their stronger competitors. These conditions led investors to avoid anything remotely resembling telecom; the resulting capital squeeze further hurt the remaining survivors. This was not just a low point in the business cycle; it was economic metastasis, an epic failure, a full-bore meltdown.

Analysts, reporters, and other pundits have frequently sought to identify the cause of the telecom industry crash. So have industry participants them-selves, both market participants and their regulators, eager to pin the blame on someone else. At the top of many commentators' lists is the Telecommunica-tions Act of 1996, which opened up local telephone service to competition across the United States. The Act led to the creation of a huge number of new companies, many of which went public quickly and visibly, and which failed, equally visibly, not long afterwards. The Act became law just as the boom was beginning, and in many ways set the direction of the industry, making it an obvious culprit.

But readers of mystery novels know that the most obvious "perp" is rarely the one who did it. It is simply wrong to lay the bulk of the blame for the meltdown on the shoulders of the Telecom Act. While its opening of greater competition did have both positive and negative effects on the economy, and its ambiguities were, in the long run, quite harmful, both the boom that was already building up at the time of the Act's passage, and the bust that resulted, were the result of many factors, most of which were set in motion years earlier. Among them were the AT&T divestiture of 1984, the development of fiber optics, the birth of the public Internet, the explosion in wireless telephony, and the novelty of electronic commerce. And while these factors did lead to irrational economic behavior and, eventually, huge losses, they alone can hardly be held to blame, either. Rather, they built up to a coincidence of opportunity, during a time of strong economic growth, that provided fertile ground for investors and entrepreneurs alike who, not fully understanding the dynamics of the industry, jumped in with too much capital and too little common sense. This fueled a perfect storm, a confluence of factors that fed on each other, in which the impact of the whole was far larger than the sum of its parts would otherwise have been.

The story that will follow begins with the tale of the telecommunications industry from its birth in the nineteenth century up through its greatest debacles in the early twenty-first century. It is the tale of an industry whose feisty, competitive beginnings were almost forgotten as it became a staid, regulated monopoly through much of the twentieth century. Competition was reintroduced in stages, piece by piece: Terminal equipment and leased-line transmission services first became competitive, then long distance telephone calls, wireless telephony, Internet service, and finally local data and voice services. It was this piecemeal transition from monopoly to competition, a necessary change by any rational measure, that eventually led to the meltdown. The industry's most experienced leaders were monopolists at heart, some of whom had trouble adapting to a competitive world. New participants more used to competitive industries, as well as their investors, underestimated the power of entrenched monopolies. Regulators provided inconsistent guidance, at times encouraging massive investment, but often leading to endless litigation and regulatory uncertainty, eventually helping to create a most unpleasant investment climate.

Acknowledgments

I could not have produced this book without the assistance of the many people I have worked with over the years, both co-workers and clients, who have exposed me to a wealth of information that I hope to be able to share a tiny fraction of herein. I am especially thankful to my former teammates at the late lamented Arthur D. Little Inc. and the recently revived BBN Corp., and at gone-but-not-forgotten Digital Equipment Corp. I also wish to thank the many people I worked with on various standards bodies and, more recently, the many experts whom I have been fortunate to correspond with on the Internet. Credit is also due to Google, the greatest research tool ever developed, and the many information providers on the Internet. Special thanks also go to my wife, Judy Hyatt, and my children, Ethan and Amelia, for indulging me during the preparation of this book.

1

Ma Bell and Her "Natural Monopoly," 1876–1969

The telecommunications industry has a deeply ingrained heritage of monopoly. Not that it always was one, or that it always needed to be one, but for almost a century, monopoly was far more common than competition. And it was the heritage of this monopoly that shaped the industry going in to the boom years of the 1990s. While some competition had begun to take root, it was being introduced sector by sector. Competitive suppliers were dependent on monopolies, and monopolies, to some extent, depended on competitors. So it may be helpful to determine how and why the monopoly came about in the first place.

Natural and Unnatural Monopoly

The usual explanation, which the "Bell System" as well as other monopoly telephone companies worldwide used for years, was that telecommunications was a "natural monopoly." To the average person, this phrase simply implies that a monopoly exists as sort of a force of nature, something inevitable like the weather. You may not like the weather, but you do not argue with Mother Nature. And you did not argue with Mother Bell. Regulators liked this argument and compounded it with rules and regulations designed to enforce the monopoly.

But such *de jure* monopolies are not the same as a true natural monopoly. Indeed the term *natural monopoly* has a specific meaning to economists. It is what happens when a given business has sufficient economy of scale, and a high enough entry cost, such that a new competitor would necessarily have higher unit costs than an incumbent, and have little chance of succeeding. In such a case the lowest average cost is theoretically achieved when there is only one provider [1]. Natural monopolies thus take care of themselves. They do not need protection; indeed, it is more likely that governmental action be taken to break them up, in order to let the benefits of competition happen. Or, as happened to the telephone industry, a natural monopoly is subjected to regulation as a substitute for the competition that would otherwise keep profits in check. Natural monopolies are not always safe: New technologies can create substitutes or erode their scope. The telecom boom of the 1990s occurred as the monopoly was eroding;perhaps some of the business failures that resulted were caused, in part, by underestimating the residual natural monopoly power of the incumbents.

Western Union

The first commercial form of electrical telecommunications was the Morse telegraph. It was a revolutionary invention that, along with the railroad, reshaped the United States and Europe during the 1840s. No longer were goods and messages conveyed by horse; the iron horse could now carry people and goods over long distances, whereas the telegraph could send information over long distances in minutes.

The telegraph industry in the United States had some early competition, but a number of companies came together under the control of Western Union. It held monopoly power, not by regulatory fiat but by a combination of natural monopoly (economies of scale) and aggressive business tactics. But Western Union was apparently becoming a bit complacent by the 1870s. Its digital transmission technology was about to be upstaged by another revolutionary invention, analog transmission of voice.

It is still debatable as to whether the telephone was really invented by Alexander Graham Bell or by Elisha Gray. Bell's original telephone, for which he was granted the patent in 1876, did not have a separate mouthpiece and earpiece. It probably did not work well at all. Gray is credited with inventing the microphone; his two-part telephone design was closer to what actually caught on. Legend has it that Western Union turned down a chance to buy Bell's patent because the company did not think that the telephone would ever catch on. A more likely explanation is that it thought his patent was not valid; the company had instead bet on Gray. Western Union had previously merged its own manufacturing operations into Gray's company, Gray and Barton, creating the Western Electric Company, and both Gray and Bell had filed patent applications on the telephone. But Bell had the better patent lawyer; Bell's company sued Western Union for patent infringement and settled in 1885.

Patent Protection

Bell's patent granted a monopoly on the telephone for 17 years [2]. During that time, telephone service was introduced into a number of major American cities. Switchboards were set up and telephone poles erected. The American Bell Telephone Company grew rapidly, but telephone service was simply not available in areas the company chose not to serve. American Bell adopted a vertically integrated business model: Following its acquisition of Western Electric in 1881, it was both equipment manufacturer and service provider.

Bell's original patent expired in 1893, at which point any number of new players entered the scene. Within the next 10 years, two million non-Bell telephone lines were installed across the country [3], and American Bell's market share was down to around 40%. While many of the new telephone companies served areas that Bell had ignored, others provided head-on competition in Bell Telephone's own markets. But Bell continued to collect and use a variety of patents to protect its interests.

One of Bell's main competitors for the manufacture of telephone gear was Kellogg Switchboard and Supply. Sometime before 1903, Bell secretly bought controlling ownership, through a trust, from a relative of the founder. Kellogg supplied many of the new competitive service providers. Bell brought a patent infringement suit against Kellogg and its customers, which Kellogg of course did not vigorously defend. This effort only ended when the secret ownership was revealed, and a lengthy court battle voided the sale as anticompetitive.

Another patent was more successful at helping Bell cement its grip over the industry. Telephone wires have a limited range: A plain copper wire pair can only carry intelligible voice, between conventional telephones, for a few miles. Michael Pupin, an immigrant from what is now Serbia who taught at Columbia University in New York, filed a patent in 1899 covering the use of *loading coils* on telephone lines. Noted British physicist Oliver Heaviside had proposed the idea a few years earlier; both Pupin and a Bell engineer, George Campbell, filed to patent the technique at nearly the same time. Loading coils (inductors placed in series with the line at fixed intervals) cancelled the capacitance between the two paired wires on the line, greatly reducing the voice-frequency attenuation of the line and allowing calls to go for at least tens of miles. Unlike the more famous Gray-Bell patent dispute of 1876, the patent office ruled in favor of Pupin. Bell purchased the patent from Pupin for $455,000, giving the company control of the only method then known for allowing intercity telephone calls. (Amplifiers came later, after the vacuum tube was invented.) So while local telephone service was, at that time, competitive, only the Bell Company could provide any sort of long distance service.

The Kingsbury Commitment

The telephone industry reached a turning point in 1912. Before then, AT&T (a name adopted in an 1899 corporate reorganization) had refused to interconnect its network to the thousands of independent telephone companies. Controlled at that point by famed robber baron J. P. Morgan, AT&T sought instead to purchase its competitors or otherwise use its strength to monopolize the industry. The U.S. Department of Justice filed suit, and in 1913, AT&T Vice President Nathan Kingsbury agreed to a settlement that became known as the Kingsbury Commitment. This provided for interconnection between AT&T and all of the independent telephone companies [4]. AT&T also agreed to stop buying up independents, except under special circumstances (such as bankruptcy), and it sold the controlling stake in Western Union that it had acquired in 1909.

Kingsbury marked the beginning of a new industry structure. AT&T was to remain the undisputed king of long distance and by far the largest of the local service providers, whereas independent telephone companies were allowed to carve out their own niche markets. The number of independents that actually competed with Bell declined, whereas new independents provided service to ever-more-rural territories. The 1921 Willis-Graham Act explicitly favored a regulated monopoly structure, allowing competitive local providers to merge. None of the remaining competitive independents survived the Great Depression. By the time the Communications Act of 1934 was passed, creating the Federal Communications Commission (FCC), there was no dispute that telecommunications was a regulated monopoly industry. AT&T's Bell System telephone companies controlled about four-fifth's of the country's telephone lines, whereas several thousand independents provided service to the rest, mostly in rural areas.

AT&T's monopoly only grew tighter over the years. Not only did the regulated telephone companies provide both local service and the sole access to long distance, but they also had a legal monopoly on "terminal equipment," devices such as telephone sets and switchboards that attached to the network. A narrow exception was carved out for press Wirephoto machines, based on the First Amendment's guarantee of freedom of the press. In 1949, as the national television networks were being developed, the FCC denied the broadcasters permission to own their own microwave relay networks. It essentially turned over the civilian part of the microwave spectrum to AT&T's control, forcing ABC, NBC, and CBS to interconnect their stations via AT&T's service. That mid-century period was the height of the monopoly's power. One could not compete with AT&T, and one could not even self-provision services that AT&T was willing to offer, or for that matter some services that AT&T *might be able* to offer, even if it chose not to.

The Slow Pace of Progress

As a protected monopoly, AT&T had little concern about competitors developing new technology. Its own research arm, Bell Labs, was responsible for many breakthroughs, but many of these seemed to be at cross-purposes with the parent company's apparent goal of meting out progress in carefully measured dosages.

This was apparent, for example, in the development of the dial telephone. During its original 17-year monopoly, telephone service was handled entirely via manual switchboards. AT&T employed thousands of operators, who sat in large rooms at boards full of plugs and jacks. While telephones had numbers, it was possible for an operator, especially in a small town, to place calls by name.

Almon B. Strowger was an undertaker in Kansas City who became convinced that the Southwestern Bell telephone operators were diverting his calls to a competitor. In 1888, he set out to address the problem via technology; in 1891, he was granted a patent on the first automatic dial telephone exchange. Although he sold the rights and did not long remain in the telephone industry, Strowger's design caught on among independent telephone companies. The company founded on his invention, Automatic Electric, supplied dial telephone exchanges to many of the independent telephone companies. During that early era of competitive telephone service, many cities had all-manual Bell networks competing with dial-enabled independents. AT&T itself did not begin to automate its exchanges until the 1920s. Although the Strowger system, sometimes called Step-by-Step, was a worldwide success (manufacture continued into the 1960s), the first Western Electric dial exchanges used instead a homegrown variant called Panel. Eventually, long after Strowger's patents were history, Western Electric adopted the Strowger design as well, deploying "stepper" switches in many parts of the country.

While progress and innovation are nowadays almost universally lauded as being for the better, slow innovation had a certain logic of its own in the regulatory environment of the mid-twentieth century. Slow depreciation schedules could result in lower local service rates. That was the main priority of many regulators; it was something that the average consumer could understand.

Depreciation has to take into account both the natural failure rate of old equipment and the economic impact of obsolescence. No one ever accused the old Western Electric of shoddy manufacturing; its gear was manufactured to last. Forty-year life spans for electronic equipment such as telephone switches seem almost absurd in today's fast-paced world, but they were the norm for decades. For many years, AT&T did not even selectively depreciate its capital plant based on actual life span; instead, all capital expenditures for a given year were lumped into a "vintage group" and depreciated at one rate. Thus telephone poles, switches, and cars were all depreciated together, generally at rates set by state regulators. (A second set of books was needed for tax purposes, since state regulators and the Internal Revenue Service had different depreciation schedules.)

The Smith Decision and Universal Service

One of the policy goals of the FCC's early years was to promote *universal service*, making the telephone a standard part of every American home. This had both public and private benefits. The public, of course, benefited from policies that made telephones affordable. This was, to some extent, used to justify the monopoly: Monopoly profits (i.e., prices in excess of what they would be in a competitive market) could be used to subsidize the price of service for those who otherwise might not be able to afford it.

This largesse was divided between two broad classes of recipients. *Rural* telephone service is far costlier to provision than urban; since most of the cost of service is in outside wire, the low density of rural areas leads to very high capital expenditure requirements. *Residential* subscribers in general also got a break: Business line rates were kept much higher than residence line rates, so that the former could subsidize the latter.

The mechanisms of the subsidy were anything but transparent. Besides the disparate rates for residential and business lines, AT&T charged far in excess of marginal cost for long distance service. AT&T's Long Lines Division then shared the take with the local telephone companies via the *separations and settlements* process. While the rules for this were arcane, they generally involved dividing the fixed cost of local service between interstate and intrastate jurisdictions. The interstate portion was covered via long distance settlements. The local companies' share of the long distance bill was calculated based on both the proportion of line usage that was interstate and the company's relative investment: A local telephone company with higher average costs, like most of the independents, would thus get a higher percentage of the toll—it could even exceed 100%. Urban Bell System subscribers were, in effect, paying part of rural subscribers' bills.

This system dated back to a pivotal decision by the U.S. Supreme Court in 1930, *Smith v. Illinois Bell*. That ruling held that local telephone lines were, in part, subject to federal jurisdiction because they carried a mix of interstate and intrastate traffic. The net result was to move some of the cost of local service onto the monthly toll bill. This decision has remained the law of the land. The FCC eventually decided to recover much of this fixed cost on a fixed basis, rather than from usage charges, thus resulting in the "FCC line charge" on modern phone bills.

Less widely discussed is the private benefit that AT&T received from the policy of universal service. The value of a network is in large part a function of the number of users [5]. If the telephone were a luxury, then certain tasks would have to be left to other media, such as letter post. Universal service, among other things, allowed businesses to count on their customers' having a telephone, thus making business lines more valuable. By enthusiastically embracing universal service as a policy goal, AT&T both protected its monopoly and grew its business. Regulation of its prices was an acceptable trade-off.

The Final Judgment

While the mid-century FCC was an unabashed fan of monopoly, not everyone in Washington agreed. The Department of Justice filed an antitrust suit against AT&T in 1949, demanding that the company sell Western Electric. This would at least create some competition in equipment manufacturing, which was clearly not a natural monopoly. Western Electric had until that point been involved in numerous other lines of business, such as audio equipment, broadcast radio transmitters, and hearing aids, but it was also in most cases the sole supplier of many types of equipment to the Bell System companies.

The case was tentatively settled in 1956 in a decision known as the Final Judgment [6]. AT&T was allowed to keep Western Electric, but the latter's operations were restricted to those needed to support the telephone company and the federal government. The independent telephone companies thus had to purchase their equipment from a variety of smaller companies. A few states did at various later times require their Bell Operating Companies (BOCs) to purchase some of their equipment from competitors, but that was not the norm.

Hushaphone and the First Cracks in the Monopoly

A modicum of competition survived the mid-century. Western Union, while no longer the powerhouse it once was, still maintained its own telegraph service. By that time, the Morse telegraph had long been replaced by the electromechanical teleprinter. In 1939, Western Union finally cemented its monopoly on domestic telegraphy by acquiring Postal Telegraph from ITT Corp., while divesting its international operations (an odd island of competition). A dial-up teleprinter network, Telex, was developing worldwide; Western Union joined in. AT&T had a competing service, TWX, built out of the emerging long distance dial telephone network. (Direct distance dialing was introduced to the public in 1951 and rolled out to most telephone subscribers within the next decade.) The two networks competed until 1981, when Western Union acquired TWX from AT&T. (A bankrupt Western Union sold both to AT&T in 1990.)

AT&T's view of its voice telephone monopoly, which was generally backed up by the FCC and state utility regulators alike, would have been humorous were it not tragic. The mere concept of "foreign attachment" was interpreted so broadly that even plastic telephone book covers were technically forbidden. Passive attachments such as headset shoulder rests, while not uncommon, were technically in violation of the rules. And thus the FCC ruled for AT&T against allowing the attachment of the Hushaphone—a plastic and metal cup that slipped over the mouthpiece to keep out background noise—to a telephone. But in 1956, the makers of the Hushaphone secured a court ruling that permitted their device to be attached. The basis of the decision was that the Hushaphone was "privately beneficial" without being "publicly detrimental." That was to become the new standard.

The next major opening occurred in 1959, when the FCC's *Above 890 Megacycles* decision reversed the course it had adopted a decade earlier and broadly authorized the construction of private microwave radio systems. A company with sufficient need for bandwidth could now self-provision it for its own use. AT&T then engaged in a competitive response, something that it had not regularly had to do for decades. It introduced a new tariff called TELPAK, by which large numbers of private line circuits within a single company's network could be priced as if they were built along a dedicated microwave route. (The circuits themselves were installed as before; the TELPAK customer, however, could create a fictitious route map of high-bandwidth pipes, from 12 to 240 voice-equivalent channels apiece, among it sites. TELPAK *pricing* was based on the resulting imaginary pipe mileage and the number of channel terminations along the way.) TELPAK successfully discouraged some actual competition, but it was withdrawn in the 1970s when the FCC overturned the general prohibition on sharing and reselling circuits that kept TELPAK's fictitious pipes from being, on average, particularly full.

The Disruptive Transistor

Other than a few private microwave systems, the 1960s saw little pressure on the monopoly. But all was not well: Technological progress was creating new demands on the network, as well as new sources of supply for potential competitors. Key to this was a Bell Labs invention—the transistor—that was worth far less to AT&T than to the rest of the world.

We do not think much about the humble transistor nowadays. We tend to think instead about products that contain thousands or millions of them, for example, very large scale integration (VLSI) chips such as microprocessors and memories. But these complex semiconductors trace their origins back to 1947's epic discovery that a small current properly applied to a semiconductor crystal can cause a large change in the current passing through it. The original transistors were used as substitutes for the then-ubiquitous vacuum tube, first in size-sensitive applications such as hearing aids, and later, as the price fell, in consumer products such as hand-held radios. (In consumer parlance of the 1960s, the word "transistor" was often used to refer to a transistor radio, typically a hand-held low-fidelity AM model.) But it was in the fledgling computer industry that transistors arguably had their biggest impact. Even a simple digital computer of the early 1950s needed thousands of vacuum tubes, each consuming several watts of electricity to run its filaments, thus using huge amounts of power and generating vast amounts of heat in the process. The market for room-sized tube computers was small. Transistors enabled computers to be smaller, faster, cooler, and cheaper. While few businesses owned a computer in 1955, they were commonplace by 1965.

Even in those early days, it was clear that computers needed to communicate: For one thing, users generally sat at terminals some distance from the "glass houses" where the computers were kept. (Actual computer networking as we know it today had yet to develop.) This was, on the one hand, an opportunity for the Bell System to increase its business; even in the early 1960s, forecasts had data traffic levels eventually eclipsing voice. But it was terribly disruptive as well. Computer technology changed far too rapidly for Bell's slow depreciation schedules, let alone its glacial rate at which it introduced new services. Computers and semiconductors thus threatened the monopoly business model both from outside the network, where they led to demand that could not be easily sated, and from inside, where faster-moving technology threatened to make obsolete billions of dollars of undepreciated electromechanical and tube-based equipment.

Federal policy finally began to catch up with technology in the late 1960s, when the FCC made two epic decisions that permanently changed the scene. The *Carterfone* decision ended the telephone company monopoly on terminal equipment; it led to the competitive availability of a wide range of devices, including modems, answering machines, PBX systems, and cordless phones. The *MCI* decision started the demonopolization of long distance service. While narrowly tailored, it released a freight train that could not be derailed, leading, eventually, to the inevitable breakup of the Bell System.

The pattern of monopoly that took hold in the United States was matched around the world. Many countries viewed the telephone network as a natural function of their existing postal monopolies. Others eventually nationalized privately owned telephone companies. As the monopoly was being dismantled in the United States, other countries followed, often very slowly and cautiously. First their monopolies were privatized; competition was then slowly introduced.

A newly competitive market, of course, attracts new entrants, as well as the capital that these new entrants require. AT&T stock had always been a stable, yield-oriented "widows and orphans" issue. New players raised the risk-reward factor. And with the telecommunications industry in almost constant flux, investors would often have difficulties assessing both the risks and the potential rewards. Carterfone and MCI were early stages in a long process that eventually led to the telecom boom of the 1990s and the meltdown that followed. The process has proved extremely beneficial for users, but much more of a mixed bag for investors.

Endnotes

[1]   See, for instance, *The Economist's* definition in its "Economics A-Z" glossary at http://www.economist.com/research/Economics/.

[2]   Alexander Graham Bell himself had little to do with the activities of the company that bore his name. For much of its early history, it was run by Theodore N. Vail, who, as much as anyone, was the visionary behind the Bell System. Theodore Vail's cousin, Alfred Vail, was himself a key player in the history of telecommunications; as assistant to Samual Morse, inventor of the telegraph, Vail was the actual inventor of the telegraph code that bore Morse's name.

[3]   http://www.telephonecollectors.org/singwire/kellogg.htm.

[4]   Kingsbury-era interconnection with competitors was on a toll basis; Bell subscribers could only call other Bell subscribers at the local rate.

[5]   This is the same concept sometimes expressed in the more recent aphorism called Metcalfe's Law, which holds that "the usefulness, or utility, of a network equals the square of the number of users."

[6]   The 1982 decision leading to the divestiture of the Bell Operating Companies was itself an extension of this case, hence the name "Modified Final Judgment."

2

The Rebirth of Competition

After the *Hushaphone* decision set a limit on monopoly power by its standard of private benefit without public harm, competition was no longer quite so unthinkable. But the monopoly system was deeply entrenched. State regulators in particular had cozy relationships with both the Bell System and the independent telephone companies, which numbered about 6,000 in the 1960s. Breaking even a small part of the monopoly would thus prove to be an epic battle. And as the 1960s wound down, competition opened up on two separate fronts, both of which would contribute to a major restructuring of the industry.

Carterfone Made the Network More Valuable

Tom Carter's little company did not set out to undermine the pillars of "The System." Indeed the Carterfone product itself was hardly, it seems, worth fighting over. Like the Hushaphone, the Carterfone did not make any electrical contact with the telephone network. It was an acoustic coupler, designed to allow a telephone handset to interface with a two-way radio system. Its target market was offshore oil platforms in the Gulf of Mexico—hardly the kind of mass market that threatens giants like AT&T.

Nor was Carter's the only "phone patch" on the market. Thousands of amateur radio operators had phone patches that connected their radio gear to their home phone lines. They were widely used on behalf of the U.S. military, to allow servicemen overseas to phone home, as well as to allow the ham operators themselves to make free long distance calls to places where another ham had a patch. Some of these patches were homemade; others were commercial products. Heathkit, in those days a major supplier of hobbyist electronic gear, had a phone patch kit, freely advertised in its catalog. Insofar as legality was concerned, most hams assumed that the telephone company maintained a "don't ask, don't tell" policy, but some may have also assumed that their FCC radio licenses demonstrated at least some qualification to touch a low-voltage audio circuit such as a telephone line. The Carterfone, on the other hand, shied away from even that level of connection. Yet AT&T considered it an improper "foreign attachment."

The FCC's 1968 ruling [1] echoed Hushaphone. And it went further, opening up electrical attachments as well. In the next few years, customer-provided terminal equipment could be attached to a telephone line, provided that the subscriber leased a "protective coupling arrangement" from the telephone company. Such devices rented for several dollars per month per line, making them uneconomical for simple applications such as home telephone lines, but they created a mechanism for the competitive deployment of business telephone systems, such as PBXs and key systems. One form of coupler, the data access arrangement (DAA), opened up the network to a competitive supply of modems. Of course the *Carterfone* decision itself also deregulated the use of acoustic couplers; such modems, which did not need DAAs, were very popular in the 1970s.

Registration Opened up the Floodgates

Several years after Carterfone, the FCC took the next step in opening up terminal equipment competition by removing the requirement for protective coupling devices. Terminal equipment *registration* instead allowed manufacturers to have their gear certified for direct attachment [2]. At the same time, telephone companies were ordered to adopt the new "modular" connectors; these would become the standard subscriber interface. Registration did take some time to work its magic: Testing laboratories were initially backlogged, and many devices were sold under the "grandfather clause" that waived registration for any device that any local telephone company had deployed prior to the registration deadline.

With the customer's ability to connect its own "terminal equipment" to the network, manufacturers unleashed a wave of innovation. Some existing products moved from telephone company rental items with low volumes and high prices to high-volume, low-priced necessities. The lowly answering machine, for instance, had been rented under tariff to a handful of businesses that really needed it; for example, movie theatres that used it to announce their schedules. Once customers could buy their own, they became almost a household necessity. This increased the percentage of calls that were answered, improving telephone company revenues.

Digitization from the Outside In

Does deregulation cause technological progress, or does technological progress force deregulation? This is a philosophical question with no simple answer. Technology can be applied to solve, or work around, regulatory problems. Effective regulation needs to take into account technology, in large part to make it *unnecessary* for technology to need to work around regulatory problems. Monopoly, however, removes most of the technological pressure from the equation. Absent competition, the monopolist and its regulators can choose to roll out technology at a leisurely pace. When demonopolization [3] occurs, a burst of technological progress may soon follow, as the market catches up with the possibilities that had previously been suppressed.

Such a burst of progress swept through the PBX marketplace in the 1970s. Before Carterfone, AT&T saw fit to provide its large business subscribers with switching technology that could best be described as "time-tested" and "well-proven." Right through the 1960s, a common large PBX system was the Type 701, a model little changed from the designs of the 1920s, based on Strowger's step-by-step technology. This was often accompanied by a cord switchboard. A more advanced PBX, widely deployed in the 1960s and into the mid-1970s, was the Type 770. This used crossbar technology, which AT&T had introduced to central office switching in the 1930s. Its common control logic, built out of relays, supported touch-tone dialing and enabled users to transfer their own calls; its switchboard did not use cords. During the first few post-Carterfone years, the Bell System's "electronic-type" [4] PBX systems, such as the Type 801, used wired-logic control circuits controlling electromechanical relay switching matrices. Computers were still too costly, to be sure, for all but the largest customer-premise equipment (CPE). But it was not Bell who introduced them to market.

Carterfone opened the doors on a new industry—competitive provision of telephone terminal equipment. There were, of course, no firms in the business, so it took a few years for an industry to develop. The first of these *interconnect companies* were started by entrepreneurs who found existing non-Bell products to distribute. As this was happening, manufacturers stepped up to the bat to try to create new products that would surpass Bell's designs.

Digital transmission of voice had been developed by Bell Labs in the early 1960s, beginning with the T1 transmission system. By 1975, the majority of the Bell System's short-haul interoffice transmission links were digital. T1 ran at a rate of 1.544 Mbps and carried 24 voice channels, each digitized at precisely 64,000 bps. A *channel bank* [5] at each end made the transition from analog to digital. In that year, all of the Bell System central offices were still based on analog switching technology; the flagship *1ESS* Electronic Switching System had computer control but still used a mass of reed relays to actually switch the calls.

But in 1975, digital switching entered the PBX marketplace with a bang, as new computer-controlled designed-for-interconnect digital PBX systems went on sale. A digital PBX converted the analog voice signals, from both the telephone sets and the telephone company's trunk lines, into bit streams and internally switched these streams. One of these PBX systems, the *SL-1*, was developed by Canada's Northern Telecom, a Bell Canada subsidiary that had essentially been spun off of Western Electric some years earlier. Another, the *ROLM CBX*, was introduced by ROLM Corp., which had previously been known as the manufacturer of mil-spec editions of Data General's then-popular minicomputers. The SL-1 used the same 64 Kbps digitization scheme as a T1 carrier, which had become the North American standard [6]. ROLM used a 144 Kbps scheme that was easier to implement with the semiconductors of the day, a decision that it no doubt regretted soon afterwards, once mass-produced chips implemented 64 Kbps digitization cheaply. Rolm and Northern Telecom soon challenged Western Electric for leadership in the PBX marketplace.

Other vendors jumped in too, with mixed success. Harris Corp. had some success with its Digital Telephone Systems product line, which, like ROLM, used a proprietary digitization technique. Rockwell International's Collins division built a large digital PBX system with specialized features for the automatic call distribution (ACD) market; it was popular, for example, at airline reservation centers. An Illinois start-up, Wescom Switching, built a mid-sized digital PBX, the *580DSS*, that made novel use of distributed control, dividing its tasks among several microprocessors. This turned out to be more of a programming challenge than its founders anticipated, though the 580 family found some success as an ACD. (Wescom was eventually purchased by Rockwell; its ACD business eventually evolved, under different ownership, into Coppercom, a manufacturer of small central office switches.) Japan's NEC also took a substantial market share.

By 1976, AT&T was in a hurry to come out with a PBX that could at least begin to compete with the new generation of interconnect systems. A "feature race" had broken out, with ROLM and Northern in particular racing to come out with larger feature lists. This was possible because of their computer control; new features only required new programming, which could be applied to existing machines in the field. Not prepared to offer a digital PBX of its own, Western Electric hurried to market a computer-controlled analog PBX family called *Dimension* [7]. These rapidly replaced the electromechanical PBXs in the Bell Operating Companies' lineup; although older models remained under tariff, they were no longer manufactured.

In the terminology of the day, analog-computerized PBXs were called "second generation," whereas digital ones were "third generation." By 1978, the marketplace was cluttered with new computerized-PBX vendors. Digital machines controlled the high end, though analog technology remained predominant in the smaller line sizes. Mitel, for instance, introduced a small, yet flexible analog PBX, the SX-200, that took a large market share in the under-100-line market. And distribution strategies were evolving too. Manufacturers even bought some of their distributors. ROLM, for instance, had begun mostly selling through independent interconnect dealers with exclusive territorial franchises but ended up with largely internal distribution after buying up franchisees. This provided an exit strategy for some interconnect investors.

The PBX acquisition decision largely hinged on renting versus buying. Dimension was not available for sale; it was rented, under state-by-state tariffs, by the Bell Operating Companies. In order to compete with the interconnect companies' advantage—that systems could be paid off and owned by their users, at little cost—the BOCs introduced new tariff schemes. One was called *two-tier*. It had a "Tier A," the so-called "fixed" portion, whose monthly price depended on the term of the contract and ended after the term's expiration, and a variable "Tier B," whose monthly price continued for as long as the subscriber kept the system. Two-tier was meant to mirror the separate leasing and maintenance costs of a PBX acquired under a capital lease. However, it was not a contract but rather a tariff, subject to the whims of state regulators; Bell only promised not to *ask* the state regulators to change the Tier A rates during routine rate cases. Later, after determining that many companies still trusted "the phone company" over a third party to install their PBX systems, Bell introduced a "Variable Term Payment Plan (VTPP),"[8] where the unitary monthly rate depended on the duration of the rental agreement. VTPP rates were the standard Bell PBX offering until the 1983 detariffing of PBX equipment, which also led to the end of the analog Dimension series.

The Integrated Voice and Data PBX Bubble

Digital PBX systems came to market in the mid-1970s, but what benefits did they bring their end users, when compared with analog systems? The real beneficiaries were the manufacturers. Digital semiconductor prices were falling rapidly, a trend that has continued for more than three decades and shows no sign of stopping. Manufacturers with foresight knew that a system built out of standard digital parts would become cheaper to build as time went on. By the late 1970s, large analog PBXs were already costlier to build than digital ones; the crossover point was clearly trending down. But this was not a selling point. Digital switching had the "cool" factor about it, to be sure, and that did not hurt with investors, but that goes just so far in a corporate setting.

But the late 1970s was also the time when "office automation" was a hot buzzword. Computer terminals were starting to show up on office workers' desks. Word processing moved out of the steno pool and onto secretaries' desks, and it was starting to show up on knowledge workers' desks too. More and more offices were being wired for data, which at that point usually meant 9,600 bps serial-port terminals, if not IBM-style coaxial cable-attached terminals. Computers themselves usually sat in a data center, so there needed to be a way to get there from the office.

Digital PBX vendors thus seized upon this as an opportunity to increase their wares' apparent utility. Computer data are, after all, digital. So why not pass off the digital PBX as a way to connect desktop terminals to their host computers? At first it was vaporware. Then the first integrated voice and data (IVD) PBX features began to arrive. Northern Telecom was a pioneer with its SL-1 Add-on Data Module (ADM). This attached to the side of its proprietary telephone instrument, providing an RS-232C connection for a computer terminal. The other end of the call, in the data center, typically required a shelf full of telephones, each with an ADM, next to the target computer. ROLM followed up with its own data appliqué.

But the real excitement came when start-up companies introduced PBX systems that were designed from the ground up for integrated voice and data. These promised to be the "supercontroller" of the integrated office. It was a story for customers and investors alike. Probably the most widely advertised of these was InteCom, whose start-up had been largely funded with Exxon's venture capital. InteCom promised that its feature-rich digital PBX, the IBX, would integrate voice and data for a small premium over voice alone.

The problem with most of these IVD systems was that their pricing story was as creative as their engineering story. The vendors promoted their systems as cost-effective, which was sufficient to solicit requests for proposals, but the actual prices turned out all too often to be more than a bit on the high side. Take, for example, a typical voice-only digital PBX circa 1980. The system price, installed with wiring and a typical mix of analog single-line and digital multibutton telephone instruments, was on the order of $1,000 per station. Vendors would then promote a typical IVD price of, say, $1,300 per station. So the casual reader might assume a price of, say, $300 per line equipped for data. But this was an average price among all lines. The vendors were really assuming that only 10% to 20% of desktops were actually equipped for data, and these desktops were connected, say, to half again as many ports going to the computers. So if a 1,000-line voice-only system was $1 million, a 1,000-line IVD system would be $1.3 million, but that assumed only 200 desktops equipped for data and 100 ports going to the computers. The actual price for data ports was more like $1,000, sometimes even higher!

In the early days of IVD PBXs, a cheaper alternative for sites that used the popular asynchronous ("dumb") terminals of the day such as the Digital Equipment Corp. (DEC) VT-100 family was to use a data-only switching system. These were known as *port selectors* or "data PBXs"; major vendors included Gandalf Data, Micom Systems, and Develcon Electronics. Designed to carry 9,600 bps data rather than voice, the typical price per port of a port selector was in the $300 range. IVD voice switches simply could not compete on price; worse, they also did not offer compelling feature advantages. So while almost every new PBX advertised data capability, very few ports were actually used for data.

All of this did not deter investors, of course. Digital PBX companies were, for a time, hot stocks. But the stock market in the early 1980s was more conservative then that it became during a later boom. The venture capital business, on the other hand, was less risk-averse than the stock market and was always looking for a "story." While large companies largely dominated the first round of IVD PBXs, the early 1980s saw a new round of start-ups that emphasized a new variation on the theme.

Although we take the concept of "local area network" for granted and even forget, at times, what the ubiquitous acronym LAN stands for, in those days the idea was new and hot. The first LAN, arguably, was Datapoint's ARC, which was in production in the late 1970s. But it was Ethernet that really created the industry. Xerox Corp. patented Ethernet in 1973; by 1980, Xerox, Digital Equipment Corp., and Intel Corp. had agreed on a specification, published it, and committed to its manufacture. Ethernet hardware began to hit the streets by 1982. While the earliest Ethernet boards—adapters plugged into the minicomputers of the day—cost more than $2,000 apiece, the price plummeted after several semiconductor makers created Ethernet chips. These were soon followed by inexpensive *terminal servers*, which permitted some number of terminals to share an Ethernet connection. (Although we take desktop *computers* for granted today, the big move away from terminals was still several years off.) Soon this approach became even less costly, and more flexible, than the port selector. The IVD PBX was falling behind; LAN servers came to outsell both port selectors and IVD PBXs. IBM, it should be noted, was hostile to Ethernet and instead promoted its own LAN technology, Token Ring. Of course its synchronous (3270-class) terminals required different hardware support anyway and were rarely supported by IVD PBX systems.

Clever entrepreneurs came up with a new idea, to "integrate the LAN with the PBX." This had huge venture capital appeal, as it combined two hot trends. There was only one problem: PBXs and LANs were very different, technically, and there was no obvious way to integrate them, and no visible *user* benefit from doing so!

Two well-funded start-ups did however run with the idea, which was sometimes called the "fourth generation PBX." The more spectacular flameout was called Ztel [9] Corp., based in Massachusetts, which received capital from, among others, General Electric and NCR Corp. Ztel's plan was to support Ethernet in its PBX. It invested heavily in engineering and created a prototype of its PBX, called the PNX. It also spent heavily to design a database server. Ztel spent heavily on a manufacturing facility, stocking up to produce a design that was not yet complete. The factory sat idle for months while engineers attempted to debug a prototype of the PNX (LAN integration having moved to the back burner); finally, money ran out and the company folded. (Nobody said that foolish telecom spending was unique to the 1990s.)

Ztel's contemporary was a California start-up called CXC, whose PBX was called The Rose. This was supposed to be based on a Token Ring LAN, rather than Ethernet. CXC did actually ship working PBX systems, though the LAN integration aspect was shelved. Although not a major factor in mainstream PBXs, The Rose achieved some success as an automatic call distributor, a specialty with relatively high per-line prices.

Integrated voice-data PBXs did end up with some niche markets, even as LANs came to dominate the office. They were useful for isolated locations where voice wiring existed and a LAN connection would have been hard to achieve. And they were sometimes useful for calls between sites; compatible IVD PBXs outperformed the modems of the day, preceding integrated services digital networks (ISDN). But these were a far cry from the promise of 1980.

Computer II and the Detariffing of Terminal Equipment

Before the *Carterfone* decision, telephone equipment such as PBX systems could only be rented from the local telephone company. Carterfone created a system in which the telephone company's rental PBXs competed with so-called interconnect systems, which were usually purchased by their users (though private lease funding was also available). This was not a level playing field, although both sides had advantages and disadvantages. The telcos often complained that they could not compete fairly when their rates were regulated. The interconnect companies complained that their customers often received inferior service from the telephone companies, who were both unhappy competitors and essential suppliers of telephone lines.

This all changed as a result of the FCC's 1980 *Computer II* ruling [10]. The earlier *Computer I* inquiry [11], decided in 1970 after several years' discussion, created a distinction between "communications" subject to regulation and unregulated "data processing." However, this created a gray area—"hybrid" services combining the two that were left to be handled on a case-by-case basis. This was hardly a satisfactory solution, as the FCC soon found itself facing many such cases. So in 1976 it began its *Computer II* inquiry, seeking a new boundary between regulated and unregulated activities.

The *Computer II* decision had several important aspects. It divided telecommunications into "basic" and "enhanced" services, the former subject to regulation, the latter handled more flexibly. And apropos the customer-premise equipment business, it ruled that such equipment, ranging from lowly telephone sets to the largest PBX systems, could in the future no longer be provided under tariff. Enhanced services and customer-premise equipment could only be provided by the Bell Operating Companies (and GTE, then the largest "independent" telephone company) via a "fully separate subsidiary" (FSS) subject to strict arms-length separation requirements. This was meant to put all players on an equal footing, and it was in that regard quite successful. A few years later, the *Computer III* decision [12] relaxed the FSS requirement, instituting instead a system of accounting and behavioral safeguards; the restriction against tariffs for CPE, however, remained in effect.

American Bell and the Embedded Base

Computer II set a date of January 1, 1983, for terminal equipment detariffing. After that date, AT&T's terminal equipment sales were moved out of the BOCs and into a new FSS, American Bell Inc. (ABI). AT&T celebrated by introducing its first digital PBX that very week. Initially called Dimension/AIS System 85, the first half of the name was later dropped. It was a curious introduction. Code-named "Antelope," System 85 was a digital variant of the older Dimension 2000 PBX; its design was largely completed by 1999, and it sat on the shelf for several years waiting for an opportunity to sell without a tariff [13]. The Computer II transition allowed the embedded base of older PBXs to remain under BOC tariff for an additional year.

But the BOCs were by then rather distracted. The Modified Final Judgment (MFJ), divesting AT&T of the BOCs, had been arrived at after Computer II was passed but before it had taken full effect. The 1982 settlement of the long-running *United States vs. Western Electric* case took effect on January 1, 1984, and transferred all of the embedded terminal equipment to AT&T, while giving customers an opportunity to purchase it. The Regional Bell Operating Companies (RBOCs), in becoming separate companies in 1984, collectively kept the Bell trademark; ABI was renamed AT&T Information Systems.

The RBOCs were allowed to sell CPE through their own fully separate subsidiaries, but they were starting from scratch. The RBOCs had lost their entire embedded PBX base to AT&T in the divestiture, but they did not lose all of their enterprise customers. In particular, they retained *Centrex* service, a PBX-type service delivered using central office facilities. Centrex had been the Bell System's flagship offering for large business since the 1960s; it allowed the features of large central office switches to be used for desktop phones. During the late 1970s, after Dimension's release, AT&T had tried to phase out Centrex via an "Installed Base Migration" [14] strategy. The BOCs had requested Centrex rate hikes in most states, succeeding in some. Suddenly, once the terms of divestiture had been announced, the BOCs realized that they lost their PBX base but kept Centrex, [15] so their marketing strategy underwent a 180-degree turn. Centrex was revived and again became a flagship product.

AT&T was only required to maintain existing rates for the embedded base for a brief transitional period. Some of the older, electromechanical systems were costly to maintain, so, to state it politely, AT&T used this pricing flexibility to encourage their retirement. This created a boomlet for the PBX industry, as the bulk of the remaining electromechanical and precomputer PBXs were replaced. Between 1975 and 1985, practically the entire installed base of PBXs in the United States had been refreshed.

From that time forward, PBX sales were keyed to more natural factors, such as economic growth. And small flurries of replacement activity occurred again in 1994, when the North American Numbering Plan introduced interchangeable area codes (those without a 1 or 0 as the second digit—some older PBXs, including some remaining analog Dimensions, could not accommodate the change), and again in 1999, as the "Y2K" craze struck. The PBX industry as a whole survived these disappointments, for the simple reason that there remained some need for their services. But its boom years of 1975–1985 were never repeated, and such a boom is unlikely to be repeated.

MCI's Shared Microwave Opened New Doors

If private companies could own their own microwave networks, then why could not more than one company share a network? That was the basic idea behind Microwave Communications, Inc. (MCI), founded by William Goeken in 1963 with the stated aim of building a shared microwave link between Chicago and St. Louis. Of course a shared network is what a common carrier provides, so AT&T naturally opposed this. The idea languished at the FCC until 1969, when MCI was granted permission to provide leased-line services. And with that move, the cracks in the monopoly armor began to grow wider. Shortly afterwards, MCI was granted permission to attach its lines to Bell System local telephone lines. MCI became a formidable competitor, in the courtroom as much as in the field, and the "natural monopoly" argument began to fall apart.

Private Line Competition Led to Rate Restructuring

By authorizing MCI to provide leased lines (more often, in those days, called private lines), AT&T faced its first serious common carrier competitor in decades. AT&T's private line rates in that era were not primarily based on cost. While TELPAK rates were designed to track the costs of a private microwave network, rates in general were at best fatuous, based on the historical "value of service" concept. Private lines were, after all, a threat to long distance revenue: A corporation could install a tie line between two of its sites [16] and thus make calls at a fixed monthly cost, rather than pay per call. MCI's threat to offer lower leased-line rates was therefore not limited to AT&T's existing private line base but potentially impacted toll revenues as well.

Although MCI began in the Midwest, it rapidly expanded its network. And with competition now allowed, other so-called specialized common carriers (SCCs) received authorization. A wave of investment began; building a nationwide long distance network, even using the analog microwave radio systems of the day, was not cheap. Probably the most important of the other new SCCs was the Southern Pacific Communications Company (SPCC). The Southern Pacific was one of the country's most extensive railroad networks, and like most of its competitors, it owned a private telecommunications network as well, primarily built using microwave towers along its rights-of-way. Another new telecommunications company, Datran, was authorized in 1969; it built the first microwave network aimed at providing data communications more efficiently than existing analog networks. Owned by Dallas millionaire Sam Wyly's University Computing Corp., Datran built its own facilities from Houston to Chicago [17]. But Datran failed in 1976, its backbone was added to SPCC's network, which later adopted the Sprint brand name and was spun off from the railroad. Even in those early days of competition, bankruptcy assets were a major tool for growth.

Nonetheless, AT&T could not take lightly the threat to its own leased-line business, especially the highly profitable long-haul sector. Thus the 1970s saw a series of rate restructurings. AT&T and other local telephone companies had to provide "tail circuits" to the SCCs, whose networks were less extensive and did not include local facilities. Its private line rates were highly distance-sensitive; the SCCs specialized in the long-haul middle section of intercity routes. So AT&T began to lower the cost of long-haul circuits, raising the cost of shorter ones. It introduced the "hi-lo" tariff. This provided for low mileage rates between certain designated major cities, and higher rates everywhere else. Thus a single AT&T private line between two distant but smaller towns could be priced by paying the higher rate to the nearest major city, the lower rate for the long haul, and the higher rate to the distant destination town. This more closely mirrored the cost structure of using a low-priced SCC. Faced with regulatory challenges, AT&T then replaced hi-lo with a newer mileage-based rate structure called multischedule private line (MPL). This was based on declining-block mileage rates, such that, for example, the first mile was much more expensive than the 1,001 mile. But there were three different rate schedules, depending on whether either, both, or neither end point was in a high-density city. This rate structure remained in place until divestiture. Divestiture, of course, put AT&T and its former competitors on an equal footing; inter-LATA (local access and transport area) circuits thenceforth consisted of discrete *special access* tail circuits from the local exchange carriers, connecting to long-haul carriers at a point of presence (PoP) at either end.

Execunet Gives Birth to Competitive Long Distance

The original SCC authorizations did not specifically include switched services; both message toll service (MTS, or conventional long distance) and wide area telephone service (WATS, or bulk long distance) were expected to remain monopolies. But the SCCs were allowed to provide foreign exchange (FX) service. This consisted of a leased line into a distant telephone central office, enabling its subscriber to get a local phone number from a distant city. FX numbers are useful both for avoiding outgoing toll costs and for providing an inbound virtual local presence in the targeted city. FX circuits, it should be noted, are said to have both an "open" end (the central office) and a "closed" end (the customer site).

MCI, in the mid-1970s, created an innovative service that made FX-like service more affordable than AT&T's version, at least to smaller users. It offered measured-use shared FX. This allowed its intercity bandwidth to be shared by multiple FX customers, who would pay for their FX numbers on a minute-of-use basis. So a customer could have a closed end connection into MCI's switching system, and the open end would be a distant Bell central office. The SCCs were subject to strict tariff requirements in those days, and the FCC accepted this tariff.

Having tariffed an FX service with a switched open end, MCI also filed a tariff for FX service with a switched closed end. A group of customers could thus access an FX service by dialing in to it via a local number, in effect making both ends open. This was perhaps more than a small stretch in the definition of FX, but the SCCs were, after all, expected to be innovators. The real punch line occurred when MCI allowed customers to dial in to one open end, enter a touch-tone authorization code, and then select a destination number —the other open end—that could be a local call from any location served by the network. MCI called this service Execunet.

Customers were quick to recognize it as a competitor to AT&T's WATS and MTS. Alas, so were AT&T and the FCC. They quickly called foul. MCI pointed out that the FCC had indeed approved the tariffs from which Execunet was built—the FCC simply had not realized it at the time [18]! The FCC, perhaps not amused, froze Execunet's growth; AT&T was allowed to stop connecting new local lines to MCI's network, though it was prohibited from shutting off service entirely while the issue was litigated. A multiyear debate began on the subject. AT&T argued that while MTS rates were above cost, its monopoly profits were necessary to cross-subsidize affordable local telephone service. In 1978, a court's ruling [19] held that since the FCC had not explicitly prohibited MCI from providing switched service, it could not do so retroactively. The barn door was open and the horse was far gone.

SPCC, it should be noted, had its own answer to Execunet. Before the 1978 court ruling officially recognized switched voice competition, it rolled out a service called SpeedFAX, ostensibly aimed at carrying facsimile. Its basic price unit was the page, not the minute, but it merely approximated pages using the then-standard Group II fax speed of four minutes per page. And of course it did work perfectly well for voice, a detail customers figured out for themselves[20].

Sharing and Resale Had Profound Implications for the Future

Historically, most telecommunications services had been reserved for the exclusive use of their own subscribers. There were explicit tariff provisions against sharing or resale of most services. This prevented companies from sharing a leased line, which might not be affordable by either one alone, or from setting up shared private networks out of leased lines. It also prevented companies from sharing WATS lines, which had provided an option, but a costly one, for unmetered long distance calling. From AT&T's perspective, WATS was a bet on averages: A given company would subscribe to WATS if it thought that its usage was high enough to justify it, but the average use was not so high as to become unprofitable.

In 1976, the FCC ruled that private line facilities could be shared or resold [21]. As noted earlier, this led to the demise of TELPAK, since TELPAK's economics were based on averages; if a company—say, MCI—could purchase circuits at TELPAK rates and resell them, then average TELPAK utilization would rise substantially and AT&T would, in effect, become its own worst competitor. But even without TELPAK, resale of leased lines was a vital tool for smaller switched long distance competitors. While MCI had built much of its own network and initially could not sell Execunet calls to areas where it did not have facilities, resale meant that AT&T's own leased lines could be used by a switched long distance competitor. In 1981, the FCC finalized sharing of WATS-type services too, which by that point were all measured-use anyway, albeit at a discount from retail tolls.

Resale was important for the long distance telephone industry, but it was even more important for the development of the Internet. In the 1970s, the Department of Defense's (DoD) Advanced Research Projects Agency had financed the development of the ARPAnet, a worldwide packet-switched data network that later became the core of the Internet. The ARPAnet was technically not resale; it was the DoD's and was only used for government-related purposes. (This was construed rather loosely, but data network research was considered a valid government-related activity, as were many other private and university-related activities.) Other corporations built private data networks too, but the first public data network providers, such as Telenet and Tymnet, were technically common carriers until *Computer II* deregulated them. Unlimited sharing and resale later allowed the development of multiple ARPAnet-like data networks, under different ownership, as well as the provisioning of lines between corporations. This was a prerequisite for the eventual development of the Internet itself.

The ENFIA Agreement Made Subsidies Explicit

Once switched competition was held permissible by the judiciary, the FCC had to figure out how to regulate it. AT&T and the independent local telephone companies were both threatened; the old system of separations and settlements that dated back to the *Smith* decision depended on keeping toll rates higher than cost in order to hold down local service rates. Execunet had been paying only local service rates for its open end lines, providing service only in low-cost areas such as major cities. This was derisively called "cream-skimming."

The FCC then held negotiations among the key parties to come up with a new set of rules for competitive long distance providers [22]. As a result, long distance providers could no longer use local service, under state tariffs, to provide the open ends of interstate switched long distance. Instead their connections to local carriers would be tariffed federally, under a tariff called ENFIA (Exchange Network Facilities for Interstate Access). The ENFIA tariffs included both line-side connections to the local switches, as used by Execunet, and for the first time higher-quality trunk-side connections. These introduced the use of the 950 prefix for 7-digit uniform access to long distance carriers, versus dialing a separate local number in each city. And they introduced the concept of switched access charges for long distance carriers to use local networks.

MCI's Growth Fueled by Antitrust

Antitrust laws exist to protect competition, providing an avenue by which an aggrieved competitor can seek recourse from a monopolist that violates accepted standards of competition. MCI sued AT&T in 1974, claiming that its monopoly on long distance service was a violation of antitrust law. A jury ruled in MCI's favor in 1980, awarding $600 million in actual damages, which were tripled under antitrust law. This was eventually reduced on appeal to $113 million [23], but it helped MCI's credibility and gave it crucial funding at a time when MCI's business had been consistently unprofitable.

However, antitrust victories for competitive telecommunications providers have not been consistent. To some extent, regulated companies have claimed some relief based on their monopolies' having been subject to regulatory scrutiny. But the regulatory process itself is not an exemption from antitrust. And as we shall see, the biggest antitrust settlement of them all, the one that reshaped the telecommunications industry, soon followed; while ENFIA created a regulatory framework for long distance competition, more drastic action would impact the business framework. In fact, 1984 was not only the year named in

George Orwell's prophecy, it marked the crucial restructuring of the American telecommunications industry into one in which market forces and competition, not monopoly, would predominate.

Endnotes

[1] *In the Matter of Use of the Carterfone Device in Message Toll Telephone Service*, 13 FCC 2d 420 (1968). Note that the FCC used its authority over *interstate* telephone service to override state restrictions against foreign attachments.

[2] *Proposals for New or Revised Classes of Interstate and Foreign Message Toll Telephone Service (MTS) and Wide Area Telephone Service (WATS), First Report and Order*, 56 FCC 2d 593 (1975); *Second Report and Order*, 58 FCC 2d 736 (1976). Two exceptions to registration are noted: pay phones and party lines. In the former case, customer-provided attachments could have compromised the integrity of the billing process. In the latter case, telephone instruments provided with the party line were typically set up to ring selectively and sometimes allow automatic billing of calls to the proper party. Party line subscribers were thus a long-standing exception to the right to have customer-provided terminal equipment.

[3] It is tempting to say "when deregulation occurs," but in practice it is regulation that is needed to overcome the monopoly; absent regulatory pressure, the telephone companies could have simply disconnected service to those who had the temerity to attach their own equipment to the network.

[4] See, for instance, the 1979 Bell System Practices, Section 460-000-006, *Alpha-Numeric Index, Station, Key, PBX and Private Service Systems*. This was made available on http://www.bellsystemmemorial.com/.

[5] More precisely, the T1 transmission system carried the DS-1 (digital signal level 1) rate of 1.544 Mbps, with no inherent notion of channelization, across two pairs of copper wire; repeaters were needed every 6,000 feet. The D-series channel banks at each end mapped voice into fixed channels using time division multiplexing (TDM). To this day, the de facto standard for carrying telephone lines across T1 is based on the way the D3 and D4 channel banks did so. The actual T1 transmission system has, however, been supplanted by newer technologies such as high-bit-rate digital subscriber line (HDSL), which carry DS-1; T1 remains a nickname.

[6] ITU-T Recommendation G.711 describes 64 Kbps pulse code modulation schemes for voice digitization. (ITU-T stands for International Telecommunications Union Telecommunications Standardization Sector.) The encoding used in the United States, Canada, and a few other countries is different from the one used in Europe and most other countries.

[7] There were three Dimension variants. Dimension 100, with a nominal 100-line capacity, and Dimension 400, with a nominal 400-line capacity, were essentially two packages of the same basic design. Dimension 2000, with a somewhat more powerful processor, was built out of multiple interconnected switching nodes.

[8] Because these VTPP rates were, on the whole, usually higher than the two-tier rates that they replaced, VTPP was sometimes said to mean "virtually twice the present price."

[9] No connection to the later Z-TEL, a competitive local exchange carrier. The original Ztel's name was an acronym of its four founders' names, Zannini, Tao, Epstein, and Lento.

[10]   77 FCC 2d 384 (1980).

[11]   See, for instance, Cannon, R., "The Legacy of the Federal Communications Commission's Computer Inquiries," *Federal Communications Law Journal*, Vol. 55, No. 2, March 2003, http://www.law.indiana.edu/fclj/pubs/v55/no2/cannon.pdf.

[12]   *In re Amendment of Sections 64.702 of the Commission's Rules and Regulations (Third Computer Inquiry)* and *Policy and Rules Concerning Rates for Competitive Common Carrier Services and Facilities Authorizations Thereof Communications Protocols under Section 64.702 of the Commission's Rules and Regulations, Report and Order*, CC Docket No. 85-229, 104 FCC 2d 958 and 9 (1986); also 2 FCC Rcd 3072 (1987) (Phase II Order).

[13]   The author notes that numerous co-workers at Digital Equipment Corporation had worked on Antelope prior to 1980, including an ambitious file service machine, based on a VAX computer, that never came to fruition.

[14]   Note the initials, which allude to a company that had previously pursued a similar strategy with regard to its leased base of large computers.

[15]   While the Bell Operating Companies retained the Centrex lines themselves, the telephone instruments on users' desks were part of the embedded base and transferred to AT&T. After divestiture, a new subset of the interconnect industry was created to support Centrex with the requisite CPE; these companies were often sales agents for the RBOCs and received commissions on Centrex lines.

[16]   The term "tie line" is generally applied to these leased lines between customer switching systems, such as PBX and Centrex systems.

[17]   Sam Wyly's "official" Web site includes this 1975 article about Datran's then-current plans (see http://www.samwyly.net/highway.html).

[18]   See, for instance, Krause, J., "Free Voice-over-IP," *CED Magazine*, April 2003, http://www.cedmagazine.com/ced/2003/0403/04cc.htm. Its analogy of Execunet to VoIP is apt.

[19]   *MCI Telecommunications Corp. vs. FCC*, 580 FCC 2d 590, U.S. Court of Appeals, District of Columbia Circuit (1978).

[20]   See *TELECOM Digest* posting by Ron Havens (http://www.yarchive.net/phone/sprint.html), which also notes how the Sprint name was the result of an employee contest that followed the court's *Execunet II* ruling.

[21]   FCC Docket 20097, *Resale and Shared Use*, 60 FCC 2d 261.

[22]   Borchardt, K., *The Exchange Network Facilities for Interim Access (ENFIA) Interim Settlement Agreement*, Harvard University Program on Information Resources Policy, 1979.

[23]   Kellogg, M., P. Huber, and J. Thorne, *Federal Telecommunications Law (Second Edition)*, Aspen Publishers, 1999, citing *MCI vs. AT&T*, 708 FCC 2d 1081 (7th Circuit), cert. denied, 464 U.S. 891 (1983).

3

Divestiture: Equal Access and Chinese Walls

The domestic telecommunications industry was reshaped in the 1980s by the breakup of AT&T, which created the Regional Bell Operating Companies. This led to a fully competitive long distance sector, which in turn unleashed many billions of dollars of new investment. Ironically, this restructuring, the so-called *divestiture*, came about as a result of a judicial proceeding that initially focused on equipment manufacturing. While the divestiture did indeed have a significant impact on the telecom equipment business, its primary focus was on the network. And it was, in effect, an official recognition of the fact that the "natural monopoly" had shrunk, leaving only limited sectors to monopoly control.

Vertical Integration

During the days of the monopoly, there was, for all intents and purposes, one "telephone company." In most parts of the United States it was AT&T, via its Bell Operating Companies. AT&T was vertically integrated, via its ownership of Western Electric. This distinguished it from its European counterparts, the Post, Telephone, and Telegraph agencies (PTTs), whose monopolies tended to purchase their equipment and supplies from favored national suppliers. So while the British Post Office, Deutche Bundespost, and other PTTs were monopolies to their subscribers, they benefited from some degree of competition in their own supply chains. Indeed their *monopolies* in telephone service also made them *monopsonies* (sole purchasers) for network equipment. True, there were many PTTs for a manufacturer to try to sell to. But until the 1980s, most of the larger countries saw fit to set their own network standards, with just enough variation to discourage or exclude foreign vendors, if not outright protectionist purchasing policies. Sweden's L.M. Ericsson and American-based multinational ITT did a good business selling to the PTTs in many smaller markets around the world, but Germany's Siemens, for example, had a strong home base to preserve, and a symbiotic relationship with its home PTT.

Western Electric's close relationship with AT&T was not without controversy. For much of its history, it had several nontelephone product lines. For example, many motion pictures from the first three decades of "talkies" were made using Western Electric sound equipment; the company was also a major manufacturer of radio broadcast transmitters. But the federal government brought an antitrust case against it in 1949. Settled in the 1956 consent decree known (ironically) as the "Final Judgment" (FJ), Western Electric became subject to new restrictions on its business. It could still provide services to the federal government, and could still provide equipment to its Bell System affiliates, but it was no longer allowed to compete in most other areas [1]. Its interest in Canada's Northern Electric (now Nortel Networks) was also sold.

It is a curious coincidence that the case leading to the original Final Judgment began the year that AT&T's monopoly was extended to prevent even private microwave competition and ended the year of the *Hushaphone* decision. With the AT&T monopoly at its zenith of power, its manufacturing arm was isolated from the normal ebb and flow of competitive business; the FJ simply confirmed Western Electric's role as part of a vertically integrated monopoly, whose scope, while broad, was not unlimited.

AT&T Kept Out of the Computer Industry

The computer industry was not well developed by 1956, of course; it was being revolutionized by the introduction of the transistor, itself a Bell Labs invention. But as a result of the Final Judgment, AT&T and its manufacturing arm were not allowed to participate. Teletype Corporation, another AT&T subsidiary, did for a time have a major share in the market for computer terminals, but those products were nominally designed for use as telegraph (Telex and its counterpart TWX) machines. And attempts to expand into more advanced terminals, or terminals for the IBM mainframe world, were met with legal resistance. AT&T attempted to file a tariff for the Teletype 40/4, an IBM-compatible terminal cluster, but the FCC refused, on Computer I grounds, to consider it a telecommunications service [2]. The product thus could not be marketed.

By the 1970s, the computer industry was one of the major growth engines of the American economy, and AT&T saw itself as stuck on the sidelines. It had computer skills in-house: The Western Electric 1ESS Electronic Switching

System, first commercially deployed in 1965, used a computer of Western's own design [3]. In 1969, a small team of programmers at Bell Labs put together the first Unix® operating system, which grew increasingly popular during the 1970s as it was distributed to universities where a generation of computer science students learned it. It too could not be sold as a commercial venture under the terms of the Final Judgment.

An obvious solution to this problem would have been for AT&T to divest itself of its manufacturing operations. Although the FJ had made Western Electric into a docile component of an integrated monopoly, its role in an increasingly competitive industry remained open to question. Western Electric's prices to the Bell System licensees, while generally not made public, were known to be competitive. But even its pricing structure was problematic: Since its products were not sold on the open market, the price paid by Bell System licensees [4] was essentially an internal transfer rate. It consisted of a direct cost and a markup, the latter differing based on which "product line" the item was assigned to. Public utility commissions performing cost studies sometimes questioned whether a given product was assigned to the right product line. For example, when Dimension PBX was introduced, Western assigned it to a product line that had a lower markup than the one used for earlier PBXs. Some states viewed this as anti-competitive.

Legislative and Antitrust Actions Took Shape

The Justice Department filed a new antitrust suit against AT&T in 1974, and as competition in long distance took root in the late 1970s, a number of private antitrust suits were also filed against AT&T. These suits were consolidated under Judge Harold Greene in 1978 [5]. At this point any number of issues could have been brought to the table. AT&T wanted relief from some of the FJ restrictions, but it was largely on the defensive. Its chief, John deButts, had been attempting to make a last stand against competition, but that position was collapsing on multiple fronts.

Congress had been considering its own options. In 1976, at the behest of a friend from an independent telephone company, Rep. Teno Roncalio (D-Wyo.) filed the *Consumer Communications Reform Act*, better known as the "Bell Bill." [6]. It would have declared local and long distance telephone service to be a monopoly utility, shutting down MCI and other new competitors in their tracks, and it would have returned control of terminal equipment to the states, allowing them to shut down "interconnect." The goal was ostensibly to preserve the cross-subsidies needed for universal service. But while the bill at one point had 200 co-sponsors, and AT&T had sent an army of lobbyists to push it, the bill's initial support dried up rapidly and it failed to carry. Congress became a pro-competitive force instead. A 1979 bill filed by Rep. Lionel Van Deerlin, chairman of the Communications and Power Subcommittee of the House Commerce Committee, would have separated Western Electric from AT&T, mandated interconnection for competitive carriers, and created an explicit universal service fund [7]. Although it would have voided the FJ and allowed AT&T to enter new lines of business, the company opposed it, and it was stalled in committee.

By 1981, AT&T's position appeared to be weakening, as the chairmen of both houses' Congressional committees that oversaw telecommunications were in the hands of competition advocates, Rep. Timothy Wirth (D-Colo.), who had replaced Van Deerlin, and Sen. Robert Packwood (R-Ore.). Packwood filed his own version of the Van Deerlin bill. Later in 1981, Wirth filed his own bill, which would have forced AT&T to spin off the Bell Operating Companies while remaining subject to strict limitations on its other activities, including manufacturing.

AT&T had been hoping that the new Reagan administration would be more supportive of its interests, but the administration was divided, and the antitrust case continued. AT&T then realized that the case was not going its way, nor were any of the three branches of government particularly well disposed toward it position. Facing a divided executive branch, a hostile legislature, and an activist judiciary, a negotiated settlement became its best option. AT&T and the Justice Department came up with a proposed settlement. Their plan was to void the 1956 Final Judgment in New Jersey, reopen the 1949 case, move the jurisdiction to Judge Greene's court in Washington, and file a settlement *under the laws in effect in 1949* when the case was first filed, rather than the laws in effect in 1974 when the pending antitrust case was filed. This would have weakened Judge Greene's authority, under the 1974 Tunney Act, to examine the public interest implications of the settlement and make further changes [8]. But Judge Greene was unhappy about the scheme and instead held that the parties could only enter into the settlement as a modification of the Final Judgment if they agreed to accept Tunney Act terms. And thus, in January 1982, the MFJ, breaking up the Bell System, was announced. AT&T would divest itself of its local operating companies, while keeping Western Electric and being released from the 1956 consent decree terms. Judge Greene took public comment and got to make some modest modifications of his own to the settlement in August 1982. The Supreme Court upheld it in 1983; the curtains were finally closing for good on the Bell System.

The final plan split AT&T eight ways. See Table 3.1 for a breakdown. The 22 Bell Operating Companies were rolled up into seven similar-sized Regional BOCS (RBOCs). The "Bell" trademark became the RBOCs' collective property. AT&T shareholders were to be given shares in each RBOC, unless they chose otherwise. Bell Labs itself was split up. Much of it became part of a "central services organization" owned collectively by the RBOCs, with a mix of

Table 3.1
Bell Operating Companies Prior to 1984 and the RBOCs Originally Created

Pre-1984 BOC	RBOC	Post-2000
New England Telephone	NYNEX	Verizon
New York Telephone	NYNEX	Verizon
New Jersey Bell	Bell Atlantic	Verizon
Bell of Pennsylvania	Bell Atlantic	Verizon
Diamond State Telephone (subsidiary)	Bell Atlantic	Verizon
Chesapeake & Potomac Telephone of Virginia	Bell Atlantic	Verizon
Chesapeake & Potomac Telephone of DC	Bell Atlantic	Verizon
Chesapeake & Potomac Telephone of Maryland	Bell Atlantic	Verizon
Chesapeake & Potomac Telephone of West Virginia	Bell Atlantic	Verizon
Southern Bell	BellSouth	BellSouth
South Central Bell	BellSouth	BellSouth
Ohio Bell	Ameritech	SBC
Michigan Bell	Ameritech	SBC
Indiana Bell	Ameritech	SBC
Illinois Bell	Ameritech	SBC
Wisconsin Bell	Ameritech	SBC
Southwestern Bell	SBC	SBC
Pacific Bell	Pacific Telesis	SBC
Nevada Bell (subsidiary)	Pacific Telesis	SBC
Mountain Bell	US West	Qwest
Northwestern Bell	US West	Qwest
Pacific Northwest Bell	US West	Qwest
Southern New England Telephone	(independent)	SBC
Cincinnati Bell	(independent)	(independent)

(Southern New England Telephone and Cincinnati Bell were Bell System licensees but AT&T held only a minority interest, so they were treated as independent).

responsibilities for network administration, software development, and consulting. It was initially called Bell Communications Research, or Bellcore for short. Some years later, after the RBOCs became more of each others' competitors and thus less interested in a joint venture, it was sold off and renamed Telcordia Technologies.

The Money's in Long Distance, Right?

Common knowledge in those days, or at least conventional wisdom, was that the most profitable part of the telephone industry was long distance. High toll rates had been subsidizing local service for years. So the terms of divestiture seemed to favor AT&T, which took the long distance network as well as manufacturing. The RBOCs, or "Baby Bells" as they were nicknamed, spun-off local companies, seemed to be weaker, especially given the many restrictions imposed on them. Among these rules, the RBOCs were kept out of the long distance and manufacturing markets, so they could not compete with AT&T's remaining core businesses. And they were given a new obligation: equal access. The RBOCs would have to give their subscribers a choice of long distance provider and interconnect with AT&T and the latter's competitors on equal terms.

MCI, Sprint, and various smaller start-ups had been competing for long distance; they were the key beneficiaries of divestiture. With AT&T separated from the RBOCs, all long distance companies would theoretically be on an equal footing. But given economies of scale, AT&T would still have the advantage over its smaller rivals. And it would be several years before the lucrative 800-number business would see any competition.

Divestiture created a strict separation between the local and long distance industries. But it did not quite work out as expected: The RBOCs ended up quite comfortable indeed, whereas the long distance companies engaged in cutthroat competition. The difference, as we shall see, is that the RBOCs still had a monopoly on a critical part of the network. That monopoly was a powerful shield against the turmoil that buffeted the industry over the next two decades. The RBOCs did not, as it turned out, need the most adroit management skills in order to prosper; a little bit of litigation and a lot of lobbying went a long way. As the shackles they were born with unraveled, the Baby Bells were to grow up to be the true heirs to the old Bell System legacy of monopolistic practices.

Long Distance Rate Restructuring Had Been Planned Before Divestiture

Prior to divestiture, the BOCs (and the many smaller independent telephone companies serving mostly rural areas) divided long distance revenues with AT&T Long Lines based on the *separations and settlements* process. It produced the implicit subsidies from AT&T that ENFIA tariffs produced from competitive carriers. Separations involved some clever accounting, taking into account such factors as relative per-subscriber investment in plant and the percentage of mileage carried by each company. High-cost rural companies could collect well over a 100% share of AT&T's revenues for calls to their subscribers. And it kept an ever-growing share of the local telephone network within the interstate jurisdiction, where its heavy fixed costs could be covered by usage-based billing (tolls).

The District of Columbia Circuit Court of Appeals Execunet decision in 1977 hinged on a determination that the FCC, in broadly approving private line competition, did not properly take steps to prohibit competition in the closely related MTS and WATS (retail and bulk long distance calling) markets. This left the commission with a problem; it could affirmatively determine that competition should be disallowed, or it could accept competition and come up with new rules for it. Thus in 1978, it opened the *MTS and WATS Market Structure* proceeding [9]. The ENFIA agreement that year led to a tentative détente between AT&T and GTE, the largest incumbents, and the competitive providers. (The smaller independent telephone companies were not initially required to interconnect with competitive switched long distance providers.) There was, by then, no serious chance of going back to a monopoly. In April 1980, two days after releasing its landmark final *Computer II* ruling and before divestiture was being spoken of in public, the FCC adopted a *Second Supplemental Notice* in 78-72, proposing a system of "access charges" to replace the old settlement's scheme. In a part of its subsequent *MTS and WATS Market Structure* decision [10], the FCC held that, in general, non-traffic-sensitive (NTS) costs should be covered by fixed fees. Thus the subsidy of local monthly service, a fixed cost, from tolls should be reduced, not increased. As it happened, the divestiture decree was announced before the FCC settled the matter of access charges, so the FCC set their effective date to be the date of divestiture.

This phase of the proceeding was instituted to determine, first, whether the existing methods of compensation for exchange plant used in interstate telephone service should be replaced by a tariffed access charge framework and, second, if so, what the structure of such tariffs should be. The entry of the MFJ has effectively mooted the first question [11]. It is worth noting that the same proceeding created the "exemption" from access charges for "enhanced service providers." Because dial-up data networks carried information across state lines, the FCC considered them to be within their jurisdiction, but they had been established using local numbers, with the expectation that they would be treated as local service. The FCC thus "exempted" them from access charges, and most networks and their users simply treated these modem pools as local business lines. In 1987, a rather different FCC, under Chairman Mark Fowler, opened a proceeding to remove that exemption and thereby charge originating switched access rates, at that point generally in the range of $3 to $5 per hour, for dial-up data access. This was derided as a "modem tax" and blocked in no small part due to the public uproar, which led all the way to Capitol Hill. Rep. Edward Markey (D-Mass.), at that point chairman of the relevant House subcommittee, held hearings on the subject and made it rather clear that Congress would not be happy if the plan went ahead. Thus the "ESP exemption" was preserved. This was critical to the development of public Internet service in the 1990s; had dial-up data calls been treated as "access," it is likely that the consumer Internet would never have developed as it did.

The Baby Bells would be born with a new set of rules to guide them. It should be noted, though, that this system of access charges was not created as part of divestiture per se, merely a concurrent change in the way interstate rates were structured. Had divestiture never occurred, similar rules would likely have been implemented, even if the access charges had largely remained within the Bell System family.

Birth of the Baby Bells

The AT&T divestiture took place effective January 1, 1984. This not only marked a change in stock ownership, but it also resulted in a change to the architecture of the telephone network itself. The primary distinction between the RBOCs and AT&T was competition. The monopoly that AT&T's Bell System had held during the mid-twentieth century stemmed from a recognition of the "natural monopoly." Almost every aspect of the telephone network had shown, in those days, natural monopoly aspects. (It was hard to argue that terminal equipment was a natural monopoly; Bell's justification for monopoly was more along the lines of its being a cash cow.) But by the time MCI was authorized, the natural monopoly in long distance was demonstrably weakened. (AT&T's residual market power was, however, noted by the FCC, which continued to regulated it as "dominant" for several years after divestiture, until its market share had declined substantially.) Long distance companies still depended on the local exchange carriers; those still seemed to be a natural monopoly. So the RBOCs were given that business, and confined to it by strict limitations on their activities. For their first several years, they were not even allowed to have unregulated businesses totaling more than 10% of their total revenues.

Stopping at the LATA Boundary

When the MFJ was originally announced, the spun-off Bell companies were to be prohibited from "interexchange" service. This was an overreaching limit that would, if applied literally, have broken local telephone service for most Americans. An *exchange* is a geographic area in which calls to and from are charged at a uniform rate. Nowadays the term *rate center* is more widely used; more precisely, a rate center is the single geographic point within an exchange that is used for measuring mileage. The United States has almost 20,000 exchanges. Sometimes an exchange is served by a single central office building, also known as a *wire center*. Sometimes there is more than one wire center in an exchange; sometimes a wire center serves more than one exchange. The boundaries of an exchange are filed in a state tariff. For instance, all of New York City is technically one exchange, but the much smaller city of Boston is split among 12 exchanges. It was not rational to expect calls within Boston to be carried by a long distance provider. So by the time Judge Greene announced his modifications to the settlement, a new term had entered the lexicon, the *LATA*, or local access and transport area. This was to be the scope of the RBOCs' activities. They could carry traffic within the LATA on a monopoly basis but could not carry traffic across the LATA boundary [12]. LATA boundaries were, in general, based on metropolitan areas and state lines, the former sometimes taking precedence. The FCC drew the lines before divestiture, in a process that involved some jockeying between AT&T and the still-gestating RBOCs. Small or sparsely populated states could be a single LATA, whereas larger states were divided into multiple LATAs. So, for instance, Pennsylvania was divided into six LATAs, one of which, metropolitan Philadelphia, included all of adjacent Delaware. Five of the six New England states became single LATAs. But larger New York was divided into seven LATAs, California into 11, and Texas into 17.

The LATA boundary was, in effect, a Chinese Wall, a clear boundary with only limited, tightly controlled crossing points. If a customer wanted a leased line from one LATA to another, it would have to go through an interexchange carrier (IXC), as the long distance companies were called. The IXCs would each establish at least one POP in each LATA they directly served, and the local exchange carriers (LECs) would route traffic through the POPs at either end of an interexchange circuit. In effect, every interLATA circuit or call had three legs, each with its own price, although in general the IXC would be the party to make payments to the LECs on behalf of its subscribers. This sometimes had comical effects, at least in the short term. Before divestiture, New England Telephone could provide leased lines between Salem, N.H., and Lawrence, Mass., about five miles and a local phone call apart. Afterwards, the circuit had to come from an interexchange carrier. AT&T took over the existing facilities. Its only New Hampshire civilian POP was in Manchester. Although it had one in Lawrence, a post-divestiture circuit went from one five-mile route to a local leg (within Lawrence), an inter-LATA leg (Lawrence to Manchester), and an intraLATA leg from Manchester to Salem. This was bad enough, but it took AT&T some time after divestiture to get its entire circuit inventory straight—dividing the existing network between itself and the RBOCs was by no means simple. So one such circuit ordered in 1984 was reportedly routed by way of New Jersey, because the Manchester-Lawrence links were not yet in the computer, whereas both PoPs had bandwidth available to Trenton!

But the LATA did create a useful framework for future network architecture. In addition to providing long distance companies with exclusivity on inter-LATA communications, it conveniently partitioned the country's telephone network, including insular territories, into 200 pieces, each of which could be served by a single Point of Presence (although many IXCs did have multiple PoPs in larger LATAs). The LATA also became a convenient guideline for the 1996 Telecom Act's introduction of local telephone competition.

Although RBOCs could not offer services that crossed LATA boundaries, other local exchange carriers were not bound to the same restrictions. GTE, the largest independent, still owned Sprint and was not required to divest it; however, GTE agreed in a separate settlement to operate its local exchange networks on a fully separate basis, subject to the same "Chinese walls" [13]. (GTE subsequently sold partial ownership of Sprint to United Telephone, another large rural chain, which merged its own long distance subsidiary, US Telephone, into what briefly became US Sprint. GTE sold United the last of its Sprint ownership by 2001; along the way, United renamed itself Sprint Corp.) Connecticut's Southern New England Telephone, which had been a minority-owned BOC, was not subject to MFJ restrictions and dabbled in long distance and data services. But even the independents worked within a new industry structure that was based around LATAs.

IntraLATA service was initially a monopoly of the RBOCs and other franchised (incumbent) local exchange carriers. Intrastate interLATA service was generally competitive, although AT&T retained a temporary monopoly in some states. Within a few years, though, many states had implemented intraLATA competition, and the FCC eventually ordered implementation of a second presubscribed carrier for intraLATA toll calls (which previously would have gone to the LEC).

Access Charges Milked the Monopoly

Three types of "access charges" were created concurrently with divestiture to recover the jurisdictionally interstate portion of network costs. Most consumers are familiar with the fixed monthly charge on the phone bill, initially $3.50 per month for a residential line. This is short for "customer access line charge," where "access" modifies "line"[14]. (A hyphen would have been appropriate.) It has never been a charge for access *to* the long distance network, as many consumers apparently believe, nor is this "FCC line charge," as some bills list it, a tax. It is the part of the carrier's monthly rate that recovers, subject to a degree of pooling among carriers, part of the local loop cost that is assigned to the interstate jurisdiction.

A second type of charge is for "special access." This covers leased circuits and data facilities, both when provided as the intraLATA leg of an interLATA circuit, and when provided on an intraLATA basis while subject to interstate jurisdiction. A leased line used for data within a LATA, for example, is considered jurisdictionally interstate if 10% of the traffic on it crosses state lines [15].

Internet access circuits within a LATA are thus typically tariffed as special access, not leased lines subject to intrastate jurisdiction.

The third type of access charge was "switched access," which covers the intraLATA leg of a long distance call. Switched access was our old friend ENFIA, originally created for MCI and other competitive carriers, now applied uniformly. It began as a fiendishly complicated tariff, with several components, some of which were ostensibly cost-based, some more openly intended to subsidize fixed network costs. Some of the latter components, such as the *carrier common line* and *transport interconnect* charges, were largely phased out by 2000. Charges for local switching, tandem switching, and interoffice mileage remained in effect after two decades, albeit at dramatically lower rates.

Switched access was introduced in four forms, called feature groups. *Feature Group A* referred to line-side connections to end offices; that is, an ordinary local telephone line, when used as a leg of an interstate call. The original Execunet circuits would have met this definition; it was also applied to the open end of interstate foreign exchange lines. But its use by long distance companies was already declining by divestiture. Not many customers wanted to dial a local number, wait for a second dial tone, enter an access code, and then dial their destination number! Interexchange carriers that used this were thus initially given discounts for "nonpremium" access. *Feature Group B*, created by the ENFIA agreement, was already, by 1984, becoming more common. It was a trunk-side connection to the network accessed by dialing a uniform seven-digit access code beginning with 950. Sprint, for instance, had the nationwide Feature Group B code 950-0777. *Feature Group C* referred to AT&T's specific predivestiture arrangements, remaining in place only until equal access could be instituted in a given central office. *Feature Group D* referred to equal access connection, which was phased in during the 1980s. Feature Group D had two novel features. One, *presubscription*, allowed each local subscriber to specify a carrier for "1+" long distance calls. (Before equal access, this was always AT&T, or the local exchange carrier itself for intra-LATA calls. IntraLATA presubscription did not occur until after the Telecom Act took effect.) The second was 10+ dialing to specify a carrier; for instance, Sprint's equal access code became 10777. (When *10xxx* codes ran low in the 1990s, this format was changed to *101xxxx*, and Sprint's code became 1010777.)

Switched access rates were especially advantageous to the local telephone companies, including the newly created RBOCs. In 1984, the national average rate for switched access charges was 17.26 cents [16], more than half of the average revenue per minute for international and interstate calls, which was 32 cents [17]. The RBOCs' own rates were, in general, somewhat lower, because the national average was dragged up by the high-cost independents, and the RBOCs initially had to pool their access revenues with the independents. It was clear, nonetheless, that the RBOCs were themselves the major beneficiaries of long distance revenues.

Ordinary long distance rates thus did not fall immediately upon divestiture, even though competition was theoretically boosted. Indeed some rates, for short-haul and late-night calls, generally increased because access charges were levied without regard to the time of day or the actual price or intercity distance of the call. Predivestiture toll rate matrices, with time-of-day and mileage-based pricing, eventually gave way to "postalized" long distance rates.

Selecting Equal Access Carriers

The MFJ itself was designed to break the back of AT&T's long distance monopoly. In addition to spinning off the RBOCs, it called for the RBOCs and GTE to offer subscribers a choice of primary interexchange carrier (PIC) as part of the equal access process. As exchanges were upgraded to equal access, which occurred in most central offices during the mid-1980s, subscribers were given a ballot on which to select their PIC. A customer would always be assigned to the PIC it selected. But if a customer did not return the ballot, it was randomly assigned to a carrier; each carrier's share of these customers was proportionate to its share of the returned ballots.

800-Number Competition

Divestiture and equal access did not create competition for one very important sector of the long distance industry, reverse-toll dialing (800 numbers). This remained AT&T's monopoly for several years. The first 800-number competition was limited, based on assigning specific NXX codes to interexchange carriers; for instance, 800–950 was assigned to MCI. A company thus could not move its existing 800 service to a different carrier, but it could get a new number from a competitor.

During the late 1980s, the RBOCs implemented the first phase of intelligent networks (IN), a program to allow central office switches to make use of external computers to perform certain functions. This made 800-number portability possible. By 1993, customers could move 800 numbers between carriers or request any vacant number from any carrier.

800-number portability is based on a database look-up; the originating local carrier performs a "database dip" when a caller dials a toll-free number. The nationwide database, which is accessed via the Signaling System 7 network, identifies which carrier owns that specific 800 number. The call is then handed off, and the serving carrier figures out, using its own database, how to deliver the call to its subscriber. Some low-volume 800 numbers are mapped to local telephone numbers, but high-volume lines are almost always terminated directly on the 800 carrier's own switch, typically reaching the customer via special access facilities.

The Big Three long distance companies invested heavily in database-driven 800 service, adding features such as automatic call distribution (to divide calls for a single high-volume 800 number among multiple locations) and time-of-day routing. These features were not commercially available from the switch vendors; the carriers did much of their own software development. Large corporate users of 800 service, such as airlines and insurance companies, were their target markets. This feature-driven competition required large sums of capital, but it helped limit the ability of smaller carriers to compete with the majors.

Competition led to falling prices for 800 service, leading to more demand, and eventually leading to the exhaust of the 800 code itself and the implementation of new toll-free prefix codes in the 888/877/866 (etc.) series. This was made possible by the 1994 update to the North American Numbering Plan, which made it possible to have area codes without a 1 or 0 as the middle digit. (Technically, the toll-free codes, like premium-rate 900 numbers and other "n00" codes, are not area codes, but *service access codes*, because they indicate a service and not a geographic region.)

The Centrex Revival

Centrex, the local exchange carrier service that used a central office to provide PBX-like internal switching functions to a business, had been a Bell System flagship offering into the mid-1970s, but its base had been withering away since the Installed Base Migration program to replace most Centrex systems with Dimension PBX. As noted before in Chapter 2, the postdivestiture RBOCs, with no installed base of PBXs, took a very different view of Centrex and once again made it their flagship offering. Bell Atlantic was especially dependent on Centrex revenues; its flagship account was the federal government, with millions of lines in the Washington area and beyond. NYNEX in New York and Ameritech in Chicago were also major Centrex markets. BellSouth may have been somewhat hurt by its earlier regulatory success, as its regulators allowed many of its old Centrex tariffs to be replaced with a costlier substitute called ESSX; most regulators in other regions had been less inclined to accept that predivestiture proposal.

But by 1984, Centrex had a serious technical disadvantage compared with most modern PBX systems. Almost all computer-controlled PBXs, including the analog Dimensions as well as popular digital systems such as those from ROLM, Northern Telecom, and NEC, had electronic telephone instruments that provided one-button access to common features, such as call transfer, hold, and conference calling. Even more important, they provided support for multiple line appearances on a single electronic telephone set, providing for features such as secretarial coverage of a group of users. This was possible on pre-divestiture Centrex but only by using multiline key telephone sets (line-selection buttons being known as "keys"). These typically used 25-pair or 50-pair wiring to the desk, leading to extensive relay-based control systems called key service units (KSUs) in wiring closets. Electromechanical key sets were very labor-intensive; adding or moving a line required rewiring at the KSU. Electronic sets, in contrast, needed only one to three pairs of standard telephone wire and could be reprogrammed from a system console or computer terminal, often remotely. They also provided extra features, like caller identification (if coming from another extension of the same PBX system), that Centrex could not yet provide.

Northern Telecom had a proprietary electronic telephone instrument, called the P-set (as in "proprietary"), for its DMS-100 Centrex and SL-100 PBX (the latter being basically the same product but sold to an end user instead of a common carrier). FCC regulations prohibited local exchange carriers from supporting any new service whose interface specifications had not been published ahead of time, so the P-set was eventually documented and registered. But it was still proprietary, a one-vendor solution. AT&T Technologies, as Western Electric had been renamed after divestiture, did not have a corresponding product for the 5ESS Centrex. Instead, it devoted its effort to supporting an emerging worldwide standard, Integrated Services Digital Network (ISDN), that would serve the same purpose. With a digital telephone set on a digital switch with digital trunks, the network would finally be digital from end to end.

Centrex remained an important market for the RBOCs. It provided them with an impetus to support ISDN and the digitization of the network. Unfortunately, it may have blinded some Bell executives from seeing the other possibilities made possible by a digital network.

Digital Switching Becomes the Norm

During the late 1970s, the PBX industry had moved from analog to digital switching for all but the smallest line sizes. AT&T's Dimension was a holdout, but only because it was waiting for 1983's deregulation before rolling out the 1979 Antelope. Yet at the same time, the Bell networks were themselves almost entirely analog. Western Electric was a Johnny-come-lately to digital switching in the central office, too.

All of the Bell System's central office switches were electromechanical until the mid-1960s. Their first production "electronic" switch was the 1ESS, first placed in service in 1965. This combined stored-program control with electro-mechanical reed-relay switching. New step-by-step and crossbar switches were discontinued a few years later. In the 1970s, the 2BESS was introduced; this was based on the same reed-relay design as the 1ESS but in a smaller package aimed at suburban and rural communities. The smaller rural-market but still analog 3ESS was introduced in the late 1970s, although very few were ever installed; by then, Nortel's small digital DMS-10 was an obviously better choice. The entire 1ESS base was also updated with a new CPU, becoming the 1AESS, which continued to be installed into 1983.

Western Electric's first digital switch was the 4ESS, a large toll switch first introduced in 1976. This became the workhorse of the Long Lines division; the RBOCs also acquired a number of them after divestiture as tandem switches. But end office digital switching in the Bell System began with foreign switches. Most prominent was the Northern Telecom DMS family. The DMS-10, a small digital switch said to be loosely derived from the pioneering SL-1 PBX architecture, won rapid acceptance in smaller line size applications (up to several thousand lines). Introduced in Canada in the late 1970s, it entered the United States in 1981; although most U.S. sales were to independent telephone companies such as GTE, it also penetrated the Bell System. The DMS-100, a large digital switch, was introduced in 1979. The predivestiture Bell System bought a few, but once divestiture occurred, it was a hot item. Western Electric first introduced its digital end office ("Class 5") switch, the 5ESS, in 1980, but the initial design had problems that led volume shipments to be delayed until 1982. The newly divested RBOCs began to replace their analog switches with digital ones and divided the bulk of their switching business between the 5ESS and DMS-100.

Not that this digitization happened very rapidly. Depreciation schedules were slow, because slow depreciation was seen to help hold down local telephone rates. Some old central office switches had been in place for 40 years or more; 1ESS-class analog switches were designed for a 40-year service life, with typical depreciation lives in the 20-year range [18]. Digital switches resembled computer equipment, whose service life was far shorter. But they also had lower operating expenses than analog switches and were more suited to the rapidly digitizing interoffice network.

The industry's move to computer-controlled digital switching took its toll in casualties among smaller switch manufacturers. Before divestiture, the 18% of local lines that were not served by Bell System companies could not use Western Electric hardware; the 1956 Final Judgment left the independent market to others. The market for electromechanical switching had dried up by the 1970s; these vendors had to create new products or exit the market. This proved to be quite difficult. The bulk of the cost of a digital switch is in software development. A larger vendor can spread this among more customers. Thus a company that prospered selling electromechanical switches to, say, 1% of the market, could not afford to spend as much on software development as Northern Telecom or AT&T/Lucent. Economies of scale became far more compelling. This was especially true given the way switch software was developed in the 1970s and 1980s. Many switches had proprietary minicomputer processors; some even used bespoke programming languages. Standard computer operating systems could not be used; telephone switch "generics" were, in effect, huge masses of embedded code built to exacting standards of reliability.

GTE, with an 8% share of local lines, had built most of its own switches at its Automatic Electric subsidiary; its GTD-5 digital switch became widespread in GTE areas, with modest sales to other independents. But it did not prove very competitive; in 1989, GTE sold partial interest in Automatic Electric to AT&T, renaming the joint venture AG Communications Systems. This became part of Lucent, which became its sole owner in early 2004, merging it out of existence after more than a century.

TRW Vidar, which had been for a time affiliated with independent LEC Continental Telephone, introduced the ITS-4/5, the first digital central office switch built in the United States. TRW shut it down in 1988; its assets, in the form of a new company called American Digital Switching, were acquired by 33 telephone companies that owned its switches and needed continuing support. That company was unsuccessful in introducing a planned new switch, the Centura 2000, and faded out by 2000. Stromberg-Carlson, an old-line manufacturer of electromechanical switches for rural carriers, was owned by defense contractor General Dynamics when it brought out its DCO, which proved fairly popular in the early 1980s. Later acquired by Siemens, the DCO was largely phased out in favor of Siemens' mainstream European digital switch, the EWSD. (The EWSD was the rare European import capable of meeting American LEC requirements.) ITT Corp., once a global powerhouse whose North Electric Company sold switches to many independents, exited the telecommunications equipment business in the early 1980s when it sold its properties to French giant Alcatel. It too had been having problems making the software for its new "System 12" digital switches competitive in the American market. Alcatel continued ITT's customer-premise equipment business under the Cortelco name but exited the American central office space until some years later, when it acquired DSC Communications.

By the early 1990s, then, the wireline central office switch business had been narrowed down considerably. AT&T Technologies and Northern Telecom split the lion's share. Siemens was the token third vendor for most RBOCs, with Sweden's L.M. Ericsson having a tiny presence in a few states. (Ericsson, however, was far more successful selling to cellular providers.) Two 1980s-era start-ups had established niche markets for themselves. Digital Switch Corp., later called DSC Communications, was strong in the long distance field, selling primarily to IXCs. Redcom Laboratories specialized in smaller, ruggedized switches, which were sold to both military applications and the smallest rural telephone companies, especially in difficult environments such as the Arctic and tropics. But a wave of new switch vendors did not arise until the Telecom Act created new local competitors.

Digitization of the Transmission Network

During the 1970s, the bulk of the AT&T Long Lines network consisted of analog microwave radio links, with analog coaxial cable along high-traffic routes. But short-haul links, those under 50 miles, were most often carried over T1 digital carrier systems. T1 carrier was the original digital transmission system. It operated at a speed of 1.544 Mbps and ran over ordinary twisted-pair copper cable. T1 carrier required a repeater every 6,000 feet, which was the same spacing used for putting loading coils on long local loops, making it easy to deploy. At the end of each T1 carrier system was a channel bank (multiplexor), which converted it to analog in order to meet the still-analog switches. Analog was mapped onto digital at a rate of 64,000 bps per channel; the mapping used by the Western Electric D3 channel bank essentially became the North American standard for digital telecommunications. Indeed that was how most of the day's standards were set, by AT&T fiat. It was not until divestiture that a formal American telecommunications standards body, Committee T1 [19], was established.

Come divestiture, the short-haul digital links became the backbone of the RBOC intraLATA networks. A number of new 4ESS digital switches were installed by the RBOCs in major markets to function as access tandems; Nortel's DMS-200 was also a popular digital tandem. Digital switches used T1 ports for their trunks, rather than analog, reducing the need for channel banks and thereby simplifying network operations. AT&T's Long Lines division, renamed AT&T Communications after divestiture, had replaced most of its analog switches with the 4ESS. Most customer owned PBX systems, other than the installed base of AT&T Dimensions, were digital. The last major bastion of analog switching was the end office, and that was gradually changing. One major impetus behind that change was the need to modernize Centrex by adding ISDN.

The First Fiber-Optic Boom

Optical fiber had been known for many years, but it was not always a very effective conductor of light over long distances. It was useful for specialty applications like desk lamps and flexible point-illumination devices, but it originally had no applications in telecommunications. Corning Glass began work in 1967 on reducing impurities in glass, which led to the development of fiber-optic communications systems. A number of manufacturers became involved in similar developments, working on problems such as the appropriate format of the fiber itself and creating long-life semiconductor lasers to drive the system [20].

In 1977, GTE installed the first such system in practical network use, carrying 6 Mbps between Long Beach and Artesia, Calif. [21], followed shortly by a Bell System link within Chicago. By the early 1980s, commercial systems were appearing in local and long distance networks, as well as undersea cables. Virtually all were digital: While optical fiber could be used for both analog and digital applications [22], it was clear by then that the telecommunications industry needed digital transmission systems.

The RBOCs were fairly quick to complete the upgrade of their metropolitan networks to fiber optics. (Analog microwave radio systems remained in some rural areas well into the 1990s, and some rustic locations may remain on copper or microwave radio links, but the latter have since been digitized.) These early fiber-optic systems were, however, vendor-proprietary. If one end of a fiber-optic cable was fed by an AT&T LTMA fiber-optic terminal, then the other end had to be an LTMA as well. Fujitsu, Northern Telecom, and others raced each other to improve the capacity of their systems. Many of the first links went at the DS3 rate (44.736 Mbps), or at multiples of DS3 such as 135 Mbps. Many of the first fibers pulled were of the *multimode* variety, whose light-conducting cores were 50 to 100 microns in diameter. But by the mid-1980s, *single-mode* cable, with a core diameter under 10 microns, became the norm for most applications. Multimode fiber was easier to connect to but it had a shorter range, which declined with the speed at which it was run. Single-mode fiber had virtually no physical limit on the maximum speed at which it could operate; its range was determined by the purity and chemical composition of the glass. Distances of 30 miles between repeaters were commonplace, enabling most links within the RBOC networks to be accomplished without repeaters. (The most recent fiber-optic cables, developed in the late 1990s, are capable of carrying high-speed signals for hundreds of miles without repeaters.)

The long distance networks also adopted fiber optics, but not all did so as quickly. There were many interexchange carriers in place for the beginning of equal access, but most were actually resellers, either with their own switching attached to other carriers' bandwidth or switchless resellers of a larger carrier. [23] AT&T began, of course, with the huge market share resulting from its past monopoly. MCI and Sprint were the largest competitors, whereas several other companies, such as ITT Corp. and Metromedia, had built smaller regional or national networks of their own. The Williams Companies, a natural gas pipeline firm from Oklahoma, created a "carrier's carrier," WilTel, to sell wholesale bandwidth at DS3 and above, using empty pipelines as fiber- optic conduits.

MCI used substantial amounts of WilTel fiber and also had deals with railroads to lay fiber along their rights-of-way. AT&T had a patchwork of rights-of-way inherited from the Bell System days, when local and long distance were not cleanly separated. It laid in fiber, providing the company with ample bandwidth between major points of presence. Sprint's original microwave network was entirely replaced by fiber; a microwave tower was even ceremonially demolished for a television commercial. Of course Sprint's all-fiber network did not go to nearly as many places as AT&T's; the latter depended on microwave for some of its thinner routes. And while AT&T was able to modernize its network and remain competitive, the rapid pace of fiber-optic deployment caused it to take a huge write-off for its undepreciated (and historically underdepreciated) analog plant.

With fiber-optic networks in place, the telecommunications industry was facing a conundrum. Before fiber, bandwidth had been scarce; this made it easy to hold long distance rates high. Incremental bandwidth requirements often required incremental capital expenditure. With fiber in the ground, however, the *incremental* cost of bandwidth was extremely low. A company might need more electronics in order to light spare strands of fiber or upgrade the speed of an already-lit strand, but this expenditure paled in contrast to the cost of laying the fiber into the ground in the first place. The industry now needed to generate new demand. Voice alone would not suffice.

This first round of fiber-optic deployment depended on proprietary gear, whose lack of interoperability primarily benefited its manufacturers. The RBOCs recognized this as a problem. One of the first projects Bellcore undertook after it was created in 1984 was to develop a standard for fiber-optic transmission systems. Bellcore named this SONET, for Synchronous Optical Network [24]. This went through a multiyear standardization process, resulting in both a domestic SONET standard and a related, but not quite identical, international (ITU-T) standard called SDH, for synchronous digital hierarchy. The bottom speed of the SONET hierarchy, STS-1, was defined as 51.84 Mbps; higher STS rates were defined as integral multiples. Optical transmission systems were named OC-n, where the n represented the STS-equivalent data rate. STS-1 was designed to comfortably encapsulate the older DS3 digital streams, each of which encapsulated 28 DS-1 streams, the original T1-carrier rate.

SONET transmission products were finally available for deployment in 1991, with the OC-48 transmission rate (2.488 Gbps) being particularly popular. AT&T had already built out most of its network using prestandard gear, though future deployments were SONET. MCI, WilTel, and Sprint were, however, large SONET customers, as were the Bell companies themselves. By the mid-1990s, most prestandard fiber-optic transmission gear had been replaced by SONET.

ISDN Fails to Make a Dent

Discussions of ISDN tend to bring out strong feelings among telecom and network professionals. Coinciding almost perfectly with the postdivestiture era,

Integrated Services Digital Network was an ambitious worldwide program to evolve the public-switched telephone network from analog to digital. And while ISDN was quite successful in many other countries, it never quite caught on in the United States. Therein lies a tale of its own.

It is hard to describe, in simple terms, just what ISDN is, because it has meant so many different things to so many people [25]. The telephone network had been built out of analog subsystems, which were being replaced by digital ones. Before ISDN, digital switches and digital transmission facilities were interconnected using protocols, such as multifrequency tone signaling, designed for the analog world: Digital islands in an analog sea. ISDN was to be the digital glue to bind them together, removing the obsolete analog artifacts, with digital local loops to extend this network to the subscriber.

ISDN development began in 1978 at the CCITT (this stands for the International Telephone and Telegraph Consultative Committee, with the initials coming from the French), which was then the International Telecommunications Union's worldwide telecom standardization branch. (It is now known as the ITU Telecommunications Standardization Sector, or ITU-T for short.) CCITT projects ("study questions") were planned based on four-year study periods, each of which culminated in a new set of published standards [26]. The CCITT began a serious study of ISDN in the 1980–1984 time frame, culminating in a preliminary set of draft standards in the 1984 "Red Book" and a much more developed set of standards in the 1988 "Blue Book." ISDN was especially important to the European telephone monopolies of the day, most of whom at that time were still known as the PTTs. Because each country's analog network had evolved separately, with dissimilar standards, equipment had to be built to a given national standard. This was particularly troublesome at the subscriber end. In an era of increasing free trade, especially among the European states, ISDN promised a single worldwide digital telecom standard, and thus "terminal portability," as well as a more globalized market for network gear.

ISDN also promised to integrate both voice and data onto one network, saving on capital and operational costs. During the late 1970s, many of the PTTs had invested heavily in packet-switched data networks based on CCITT Recommendation [27] X.25. Designed primarily for terminal-to-host connectivity using low-speed modems and analog transmission, X.25 was not nearly as popular in the United States as it was elsewhere. Some PTTs also built circuit-switched data networks, which were faster than the day's modems, based on Recommendation X.21. ISDN could provide circuit-switched and packet-switched data, as well as voice telephone service. Many network operators considered this to be compelling.

The American perspective was a bit different. For one thing, the market dynamics were different. In Europe, each PTT was both a monopoly for telephone service and a monopsony for telephone network components. If British

Telecom wanted GEC or Plessey to build to a precisely given specification, they had little choice. (The CCITT specifications had room for some national variation, necessary because of historical differences between national networks.) Divestiture, however, resulted in seven large RBOCs and numerous independents facing two major switch vendors, AT&T and Northern Telecom. This gave the latter companies more market power. Another issue was the RBOCs' rather slow perceptions. The integrated voice-data PBX boom had fizzled out prior to divestiture; the idea of using the telephone system as a local area data network was as dated as last month's lunchmeat. Yet the predominant RBOC view of ISDN seemed to be that it was designed to offer integrated voice and data (IVD) desktop connections. ISDN's lower-speed connection, the basic rate interface (BRI), had two 64 Kbps bearer (B) channels. Either one could be used for voice or data. But in the common American viewpoint of the day, it was one voice, one data—just like the IVD PBXs with which Centrex was competing. ISDN's higher-speed connection, the primary rate interface (PRI), ran over good old T1 facilities and typically offered 23 B channels (plus the signaling "D" channel) [28]. It made a good PBX trunk, as well as a useful link for higher-speed applications such as video teleconferencing. But while most PBX trunks in Europe were converted to PRI by the mid-1990s, analog and non-ISDN channelized T1 trunks remained the domestic norm.

To a considerable extent, this was driven by the RBOCs' newfound devotion to Centrex. ISDN was just what Centrex needed, at least in their eyes. It allowed for standardized multibutton feature-rich telephone sets, to compete with PBX systems' proprietary instruments, and it offered integrated voice and data to the desktop. If ISDN had lived up to its early promise, then by 1988 or so, Centrex would have been feature-competitive with the digital PBX systems of 1980! In fact, while domestic ISDN trials began in 1987, Centrex-driven rollout was slow, though a number of installations were made by the end of the 1980s. RBOC marketing material featured IVD and local area data applications. Non-Centrex ISDN was even slower to appear. When, by the early 1990s, most major European and Japanese central offices had been upgraded to ISDN and service was readily available at little price premium over analog rates, the RBOCs were dragging their heels, typically treating non-Centrex ISDN as a super-premium service.

The early ISDN push spawned a minor industry of its own, but vendors faced a more difficult road than most expected. Bellcore was actively promoting ISDN, albeit with little marketing savvy. Dozens of companies introduced ISDN terminal equipment, not to mention the extensive test equipment needed to actually make it work: Although ISDN was conceived as a standardization project, the American implementation was somewhat different. AT&T created its own ISDN specification for the 5ESS, known as AT&T Custom. Northern Telecom created its own specification for the DMS-100, known as Nortel

Custom. Needless to say, these were not nearly compatible; terminal vendors usually had to implement both formats. (There were, after all, more RBOCs than equipment vendors.) Bellcore then dragged AT&T and Northern Telecom representatives together to hash out a unified BRI specification, called National ISDN 1. This forced terminal equipment vendors to implement a *third* dialect—NI-1 was not as feature-rich as AT&T Custom, so it mainly displaced Northern Telecom's custom flavor as the latter switches were upgraded [29]. Later, NI-2 added a common dialect for PRI.

A considerable variety of ISDN products was introduced besides a variety of ISDN telephone sets. Videoconferencing systems migrated from an earlier "Switched 56" data service to ISDN. At least one company created a remote call agent's instrument, allowing its user to be part of his employer's automatic call distributor while connected via an ISDN line at a different location. (Many years later, this was suggested as a good example of a Voice over IP application.) Data gear was widespread. ISDN-to-Ethernet bridges were made by various companies, largely aimed at telecommuters: The Transmission Control Protocol/Internet Protocol (TCP/IP) network protocol suite was not yet dominant, so bridges, operating at a lower layer, could be used with Novell IPX, DECnet, NETBIOS, or TCP/IP. And a range of ISDN plug-in cards was made for desktop computers. In Europe, a German-led effort produced the Common ISDN Application Protocol Interface (CAPI [30]) standard to allow programs on personal computers to use ISDN cards without regard for underlying variations in the national protocol. This was never very popular in the United States, where the TCP/IP protocol suite was becoming standardized and PC-to-PC connections were of little interest; domestic-market ISDN card vendors instead produced drivers to allow their cards to be used with popular protocol stacks.

ISDN's difficulties in the United States were not helped by other misfeatures that made it difficult to use. For instance, the National ISDN-1 standard required that every ISDN BRI device identify itself to the network with the line's service profile identifier (SPID). The SPID was designed to help telephone companies keep control over Centrex extensions using ISDN's passive bus option, which allowed up to seven terminals to share one BRI line. But the result was that all ISDN phones or data adapters needed telephone company-controlled passwords in order to draw the ISDN equivalent of dial tone! This led to countless repair calls and installation delays.

Scott Adams, best known for his *Dilbert* comic strip, worked on ISDN strategy at Pacific Bell early in his cartooning career. In his 1997 book *The Dilbert Future*, written just before the widespread deployment of asymmetric digital subscriber line (ADSL) and cable modems, he explained ISDN's failure:

> The only thing that could limit its success was complete incompetence on the part of all the phone companies, colossal stupidity by every single ISDN

hardware vendor, and complete idiocy on the part of the regulatory over-sight bodies.

It was obvious to me that ISDN was doomed [31].

Digital Access Held Hostage to Local Measured Service

The RBOCs were primarily interested in ISDN's voice capabilities for Centrex. But its non-Centrex data capabilities were more empowering. When ISDN trials began in the late 1980s, dial-up modem speeds were typically no higher than 14.4 kbps. Compared with this, ISDN's 64 kbps was lightning-fast; with two B channels, callers could achieve 128 kbps. This paled next to Ethernet's 10 Mbps (higher-speed Ethernet was to come later), so it was not competitive in the local area network, but it was a fine substitute for a modem. And from the carrier's perspective, it used virtually the same resources, because each telephone call took the same 64 kbps worth of network bandwidth anyway. ISDN just replaced the line interface and left out the analog-to-digital converter. It did not cost the carrier any more to carry an ISDN data call than a voice call, except that in some cases, because of limitations in older equipment, clear-channel 64k service between switches had to be carried over separate trunks [32]. Pricing was a different question! Although most European PTTs treated all ISDN circuit-switched calls, either voice or data, at parity with analog calls, most RBOCs viewed ISDN as a premium service, subject to higher usage charges. In particular, almost all residential analog telephone service was offered at a flat rate (no usage charges for local calls), a situation that the RBOCs were not necessarily happy with but which regulators and, in some states legislatures, insisted upon. ISDN was seen as a way to impose measured usage on residential subscribers. So RBOCs often offered residential ISDN at measured rates, whereas analog lines could use modems at flat rates. This was a powerful disincentive to ISDN and sent a clear message to potential users. ISDN calls *within* a Centrex system—which could often be deployed on a LATA-wide basis—carried no usage charge. Some companies did create ISDN centrex groups for telecommuters and some Internet service providers did make use of Centrex and similar specialized services. But ISDN was usually not positioned as a substitute for plain old telephone service.

Not all states accepted this. California, with its large technology community, struck a compromise; Pacific Bell's residential ISDN usage was metered during the business day, not at night. Tennessee ordered BellSouth to provide unmetered ISDN at a very reasonable rate (under $35 per month), roughly half the unmetered rate in other BellSouth states, on a statewide basis. For a time in the early 1990s, before the Internet was available to the public, that state had its own ISDN subculture, with dozens of bulletin board systems using ISDN. Some of the NYNEX and GTE service areas had unmeasured ISDN "voice" usage and measured "data" usage. Since the intraLATA network almost always carried voice in a manner that could carry data on a 56 kbps basis (i.e., 7 out of 8 bits), "data over voice bearer service" became a popular option supported by many equipment vendors. This may, in fact, have been part of the inspiration for the "56k" modems that followed.

With Internet access the "killer application" for ISDN, though, less-costly ADSL services took away market share, and ISDN for Internet access eventually went into decline. Thus analog plain old telephone service remained predominant in the United States, even after ISDN became relatively ubiquitous in much of Europe and Asia.

Broadband ISDN Led to the ATM Boomlet

As early as 1985, the CCITT recognized that ISDN's basic rate and primary rate interfaces, over copper loops, were not sufficient for the long term. Copper was becoming obsolete; new networks would be all fiber optic, and standards would be needed. Thus began the Broadband ISDN (B-ISDN) program. Its goal was to create a standard for business and residential service at speeds high enough to support high-definition television (HDTV), which at that point was still not developed and whose actual bandwidth requirements were not yet known.

A few decisions about B-ISDN were made early. The initially defined access speeds were set as 155.52 Mbps and 622.04 Mbps, SONET's STS-3 and STS-12 rates, respectively [33]. Rather than have fixed channelization, a new "fast-packet" technology would be developed for it. Initially called Asynchronous Time Division Multiplexing, it was soon shortened to Asynchronous Transfer Mode (ATM).

ATM was not like X.25, Internet Protocol (IP), or other "Layer 3" packet-switching protocols. It was designed for high-speed implementation in hardware, basically at the physical layer. Thus it had a fixed-length block size called a *cell*, rather than a packet, and it did not perform payload error detection or correction in the network. Early consensus seemed to be forming for a cell payload size of 64 octets [34], with a five-octet header. But an earlier French experimental cell-relaying network had used 16-octet cells; France led a move to make the cell payload only 32 octets, which would have reduced voice packetization delay at the expense of greater overhead. Both factions wanted the cell payload size to be a power of two. An eventual compromise was reached; before the 1988 Blue Book gave the first outline of B-ISDN, the ATM payload size became 48 octets, for a 53-octet total cell.

The B-ISDN program was initially dominated by telecommunications carriers that were looking for a standard for fiber-optic networks that could compete with cable television. While cable networks theoretically could compete with telephone companies, in the 1980s the threat was indeed little more than theoretical. But it seemed like the best defense was a good offense, however slowly it was expected to happen. As a telecom-industry project, B-ISDN was defined to work in the wide area, eventually expected to become a worldwide network, but it also had the bandwidth to be competitive in the local area. So ATM LANs were very much a possibility, with a PBX-like topology.

But as the 1990s began, manufacturers had other ideas. The first practical implementations of B-ISDN were not "carrier-grade" equipment but enterprise-grade LAN switches. And they did not use SONET. The first such switch was marketed by Fore Systems, a Pittsburgh start-up (later acquired by Marconi); it used 100-Mbps fiber- optic interfaces. Other companies jumped into the fray, and the ATM Forum began as a new standards body independent of the ITU-T. In 1991 and 1992, ATM—the B-ISDN name was seen as marketing poison and largely forgotten outside of the ITU—became a major draw for venture capital. Dozens of companies produced ATM switches, ATM chips, and ATM network interface cards for PCs. The ATM Forum had a huge controversy over a standard for twisted-pair ATM for the desktop, leading to an IBM-led spin-off to promote a 25-Mbps standard that competed with the ATM Forum's 50-Mbps rate. ATM was seen, for a time, as the future of networking.

Except that it always lacked one important thing—customers. ATM's appeal to the LAN was that it was faster than 10-Mbps shared-medium Ethernet. But the Ethernet community did not stand still. Manufacturers (notably Cisco Systems) promoted *Switched Ethernet*, which took out the cable sharing. Then *Fast Ethernet* came out, at 100 Mbps, at a lower price than ATM. This turned out to be the death of ATM LANs. Although ATM could carry voice, data, or video, and assign flexible bandwidth and quality of service (QoS) to each user, most local area data users did not *need* those telecom-market features and were happy with cheaper Ethernet. ATM was pronounced dead, and B-ISDN did not resuscitate it.

But ATM did not really die. It may have gone into hibernation, but it made a comeback as a backbone for Internet service providers, and then as part of the underlying technology in digital subscriber line services. A lot of ATM companies, however, did not survive.

During the early B-ISDN era, another technology was posited as a short-term alternative. *Frame Relay* was, like X.25 and IP, based on variable-length frames; indeed it was created by taking X.25 and throwing away any function that could be performed outside of, rather than within, the network, and then some. The result was a lean, mean, simple data communications protocol, operating within so-called Layer 2, that offered some QoS options while being designed to operate comfortably up to speeds of about 2 Mbps. It was also designed to interwork easily with ATM. Frame Relay was designed to be a telephone company service—its working title began as "ISDN New Packet Mode Bearer Service," but it was introduced to market by enterprise-network and long distance vendors. Wiltel brought its Wilpak service on line in 1991, using enterprise-grade Stratacom switches that were really little more than statistical multiplexors. Startup Cascade Communications and Northern Telecom were both major vendors of Frame Relay gear, which often could also support ATM. Ironically, it was AT&T Technologies that devoted the most effort to standardizing Frame Relay, which was based on the Digital Multiplexed Interface (DMI) Mode 3 that it had designed for IVD PBXs. But AT&T did not have a successful Frame Relay product of its own until it purchased Ascend, which had earlier purchased Cascade, in 1999.

Regulators Made the Deal, but Fiber Did Not Make it Home

SONET put ample bandwidth into the RBOC backbone networks, but the "last mile," the local loop to the subscriber, was still almost always the same old twisted-pair copper wire that had been only designed to carry voice. By the mid-1980s, there was talk about "fiber to the home," which was expected to follow the deployment of fiber elsewhere in the network. By the early 1990s, the RBOCs were willing to talk about it more openly, to the point of making promises.

But they wanted something in return. The RBOCs, as monopolies, had been subject to the same kind of rate-of-return regulation that the old Bell System had, with profits capped at a percentage of their undepreciated capital *rate base*. With new technology lowering the cost of doing business (digital switches required less capital and less labor than their older counterparts), the RBOCs wanted to keep more of the savings. So they petitioned the states and the FCC to move to "rate cap" regulation. This would allow them pricing flexibility, with their rates capped at a level that reflected inflation minus a productivity factor. In exchange for this, they promised to undertake a costly network upgrade, to install "broadband" capability—usually defined then at 45 Mbps in both directions—to a large fraction of homes. Note that this was just *before* public Internet service was readily available; typical broadband applications were expected to include telecommuting, teleconferencing and "video on demand." Indeed a major goal was to provide competition for the cable companies, which at that time were largely held in low regard by customers and regulators alike, but which had won, in the 1992 Cable Act, their own freedom from most local rate regulation, in exchange for the possibility of competition.

The promises were dramatic. From the Bell Atlantic 1993 Annual Report [35],

> First, we announced our intention to lead the country in the deployment of the information highway... We will spend $11 billion over the next five years to rapidly build full-service networks capable of providing these (interactive, multimedia communications, entertainment, and information) services within the Bell Atlantic Region.

We expect Bell Atlantic's enhanced network will be ready to serve 8.75 million homes by the end of the year 2000. By the end of 1998, we plan to wire the top 20 markets... These investments will help establish Bell Atlantic as a world leader in what is clearly the high growth opportunity for the 1990s and beyond.

Likewise, in that same year's Annual Report, US West said, "In 1993 the company announced its intentions to build a 'broadband,' interactive telecommunications network... US West anticipates converting 100,000 access lines to this technology by the end of 1994, and 500,000 access lines annually beginning in 1995." These promises were typical. They were also hollow. A few telephone companies experimented with hybrid fiber-coax cable TV plant, and a very few fiber-to-the-home lines were tried, but the highest-speed residential service that the RBOCs offered by the end of the century was ADSL, using the old copper plant. Rate caps were, however, good for RBOC profits, as they substantially downsized their labor forces through layoffs and early retirement programs.

The Divestiture Era can be described as the 12-year period between 1984, when AT&T was broken up, and 1996, when the Telecom Act took effect. It began with a bang, the creation of eight companies that had formerly been one, and the creation of a fully competitive long distance market. But its net impact was, on balance, probably less than one might have expected. By the early 1990s, both the equipment and service-provider markets were consolidating. Long distance rates were falling, but the RBOCs were more profitable and powerful than had been expected. The fruits, and perils, of competition were largely yet to be seen, at least outside of the long distance arena. And the biggest round of investment in long distance facilities that was made possible by divestiture did not take place until later, when the Internet boom fueled demand that could not be met by the earlier round of fiber-optic build-outs.

Endnotes

[1] See, for example, the official history of Lucent Technologies (successor to Western Electric) at http://www.lucent.com/corpinfo/history.html.

[2] See, for instance, Office of Technology Assessment, *Information Technology R&D: Critical Trends and Issues*, 1985, p. 115.

[3] The 1ESS processor was designed for high reliability, not a characteristic of most early-1960s computers. It was in several respects quite far from the mainstream of computer design. Such custom-designed processors dominated network applications until the 1990s, when commodity microprocessors took over the role.

[4]   The "Bell System" was a group of telephone companies that paid a share of their gross revenues to AT&T as a license fee. It consisted of the 22 Bell Operating Companies that were controlled by AT&T via total or majority ownership and two companies in which AT&T held only a minority interest (Cincinnati Bell and Southern New England Telephone).

[5]   Ward, R. C., 1997 Ph.D. dissertation, *The Chaos of Convergence: A Study of the Process of Decay, Change, and Transformation Within the Telephone Policy Subsystem of the United States*, Virginia Tech, http://www.scholar.lib.vt.edu/theses/available/etd-0698-91234/, p. 265.

[6]   According to Charles Jackson, a member of the House Subcommittee staff from 1976 to 1980, Roncalio filed it at the behest of a friend from an independent telephone company but had no personal interest in the bill. As such, he filed the bill before its backers were actually ready, thus giving the opposition a "heads up."

[7]   Ward, R. C., 1997 Ph.D. dissertation, *The Chaos of Convergence: A Study of the Process of Decay, Change, and Transformation Within the Telephone Policy Subsystem of the United States*, Virginia Tech, http://www.scholar.lib.vt.edu/theses/available/etd-0698-91234/, p. 266.

[8]   Ward, R. C., 1997 Ph.D. dissertation, *The Chaos of Convergence: A Study of the Process of Decay, Change, and Transformation Within the Telephone Policy Subsystem of the United States*, Virginia Tech, http://www.scholar.lib.vt.edu/theses/available/etd-0698-91234/, p. 271.

[9]   67 FCC 2d. 757, *In the Matter of MTS and WATS Market Structure, CC Docket 78-72, Notice of Inquiry and Proposed Rulemaking.*

[10] 93 FCC 2d. 241, *In the Matter of MTS and WATS Market Structure, CC Docket 78-72, Phase I, Third Report and Order.* The docket had been opened in February 1978.

[11] 93 FCC 2d. 241, *In the Matter of MTS and WATS Market Structure, CC Docket 78-72, Phase I, Third Report and Order.* The docket had been opened in February 1978.paragraph 11.

[12] Narrow exceptions did exist: RBOCs could carry *local* calls between LATAs, though adding new interLATA local routes required FCC approval. Two "corridors" were also established from New Jersey to New York City and Philadelphia, where the existing toll facilities had been owned by the BOCs, not AT&T Long Lines.

[13] Alleman, J., and L. Cole, "Sprint—GTE's Lost Opportunity" in *The International Handbook of Telecommunications Economics*, Vol. III, Ch. 10, Madden, G. (ed.), Cheltenham, UK: Edward Elgar Publishers, 2002.

[14] Another name for the same thing is the "end user common line" charge, or EUCL.

[15] This *de minimis* test was expounded by the FCC a few years after divestiture; previously, a circuit could be considered interstate if it was "contaminated" with any interstate traffic.

[16] Federal Communications Commission, *Trends in Telephone Service*, 2002, Table 1.2.

[17] Federal Communications Commission, *Trends in Telephone Service*, 2002, Table 14.4.

[18] Depreciation for regulatory purposes was set by the states and the FCC; it did not follow the same depreciation schedules set by the Internet Revenue Service for tax purposes.

[19] The standards committee was accredited by the American National Standards Institute (ANSI), whose naming structure consisted of a letter-number combination; for example,

computer language standards were under the X3 committee. "T1" was chosen because it resembled the first two consonants in "telecommunications," not because of the carrier system of the same name. ANSI standards are named for the committee name followed by a number, such as T1.601.

[20] Hecht, J., *City of Light,* Oxford University Press, 1988.

[21] Smithsonian Institution, http://www.sil.si.edu/Exhibitions/Underwater-Web/uw-optic-05.htm.

[22] Introduced in the 1990s, the hybrid fiber-coax system used by cable television companies makes use of analog transmission of broadcast video channels over fiber optics. This allows for an inexpensive transition to coaxial cable near the subscribers' locations. Digital cable generally makes use of 6-MHz television channels within that bandwidth, within each of which a modem transmits multiplexed streams of digitized broadcast channels.

[23] Even the major carriers quietly resold each other's services; some traded bandwidth for backup purposes, while some rural LATAs were only reachable via AT&T's facilities.

[24] It was called *synchronous* because it assumed that all network elements could synchronize their timing against a single precision reference source. Earlier digital transmission schemes, such as DS3, were designed for digital islands in an analog sea, where there would be considerable variation from nominal transmission speeds. Synchronous operation made multiplexing much simpler.

[25] The author also wrote the book *ISDN in Perspective*, Reading, MA: Addison-Wesley, 1992. In a 1987 article in *Trends in Communications Management*, he described ISDN as "the invisible elephant," referring to the parable of the blind men and the elephant, and used the same phrase as the title of the book's first chapter.

[26] In 1992, along with renaming itself, the CCITT stopped publishing quadrennial "books" and instead began to release "Recommendations" whenever they were approved.

[27] CCITT/ITU-T standards are formally called "Recommendations," as they are not binding per se; this is especially important to North Americans, whose networks are generally built to separate standards.

[28] Note that Europe did not use the 1.544-Mbps DS1 rate; its equivalent was the 2.048 "E1" rate. Thus the European PRI had 30 B channels plus a D channel.

[29] NI-1 was more closely based on Northern Telecom's dialect than AT&T's.

[30] See http://www.capi.org/. The CAPI project remains active.

[31] Adams, S., *The Dilbert Future,* New York: HarperBusiness, 1997, p. 43.

[32] The original T1 specification did not permit the transmission of an octet containing eight 0's; the standard North American encoding for voice thus used only 255 out of 256 possible combinations of bits. This was corrected by the mid-1980s, with the B8ZS (bipolar 8-zero substitution) transmission technique, but most 4ESS tandems and many other digital facilities were built to the old AMI/ZCS (alternate mark inversion, zero code suppression) standard.

[33] SONET was a North American standard. The ITU-T equivalent, SDH, referred to these two speeds as STM-1 and STM-4.

[34] An octet is eight bits, grouped for transmission. A *byte* is a group of bits that typically represent the smallest addressable part of a computer's memory. A byte is usually eight bits nowadays, but historic computers used other byte sizes. It is thus more precisely correct to

follow ITU practice and describe ATM cells in octets, not bytes, though the latter term is more common.

[35]   This and many other gems were collected by Bruce Kushnick of the New Networks Institute in *The Bells' Greatest Broadband Failures*, http://www.newnetworks.com/bellbroadbandfailures.

4

The Internet Boom and the Limits to Growth

Nothing says "meltdown" quite like "Internet." For although the boom and crash cycle had many things feeding it, the Internet was at its heart. The Internet created demand for telecommunications; along the way, it helped create an expectation of demand that did not materialize. The Internet's commercialization and rapid growth led to a supply of "dotcom" vendors; that led to an expectation of customers that did not materialize. But the Internet itself was not at fault. The Internet, after all, was not one thing at all; as its name implies, it was a concatenation [1] of networks, under separate ownership, connected by an understanding that each was more valuable because it was able to connect to the others. That value was not reduced by the mere fact that many people overestimated it.

The ARPAnet Was a Seminal Research Network

The origins of the Internet are usually traced to the ARPAnet, an experimental network created by the Advanced Research Projects Agency, a unit of the U.S. Department of Defense, in conjunction with academic and commercial contractors. The ARPAnet began as a small research project in the 1960s. It was pioneering *packet-switching* technology, the sending of blocks of data between computers. The telephone network was well established and improving rapidly, though by today's standards it was rather primitive—digital transmission and switching were yet to come. But the telephone network was not well suited to the bursty nature of data.

A number of individuals and companies played a crucial role in the ARPAnet's early days [2]. It is unlikely, though, that any of them envisioned what would come of it 30 years later. The ARPAnet was designed to connect government facilities with their suppliers and academic researchers. It was not open to the public at large but grew from its original three nodes in 1969 to several hundred a decade later (see Figure 4.1). Along the way, many changes were made to the technology. This was, after all, a research network, and the network itself was the heart of the research.

In the early ARPAnet, the machines that interconnected computers were called *IMPs*, for intermachine processors. The original IMPs were built by Bolt Beranek and Newman, a Cambridge, Mass., firm that also operated the ARPAnet's Network Control Center. The network's core protocol was called NCP, for Network Control Protocol. The TCP/IP protocol suite was developed in the 1970s, and NCP support was turned off in the IMPs in 1983. TCP/IP was important because it created two separate layers with distinct functions out of what had been NCP's function. IP ran in all of the IMPs, relaying packets across the network on a *connectionless* basis. Each packet stood independent of all others, with no prior connection between the two ends required, and the full source and destination addresses in each packet. (By analogy, postcards and letters travel the same way, albeit more slowly.) TCP only ran at each end of the network, in the host computers. It resequenced received packets (IP did not guarantee delivery order), detected dropped packets, and used retransmission to correct for lost or errored packets. This allowed the intermediate systems to be relatively simple; many functions only needed to be performed at the end points.

The ARPAnet grew well into the 1980s but at one point was split into the Department of Defense's own internal MILNET and the more research- and academic-oriented ARPAnet. These interconnected networks became, in effect, the Internet, a network of networks. The ARPAnet itself was soon phased out, as the National Science Foundation took over funding of a new NSFnet backbone, and other privately operated networks, such as CSNET, became part of the Internet.

Other Packet-Switching Technologies Had Their Adherents

Although TCP/IP has become the dominant form of data networking, it did not grow up in a vacuum; there were other packet-switching technologies that competed with it or were designed for different niche markets. One has to recall that in the 1970s and into the 1980s, most end users did not have computers on their desks; they had terminals. A good deal of computer networking was about connecting terminals to their hosts. In the IBM-dominated world of

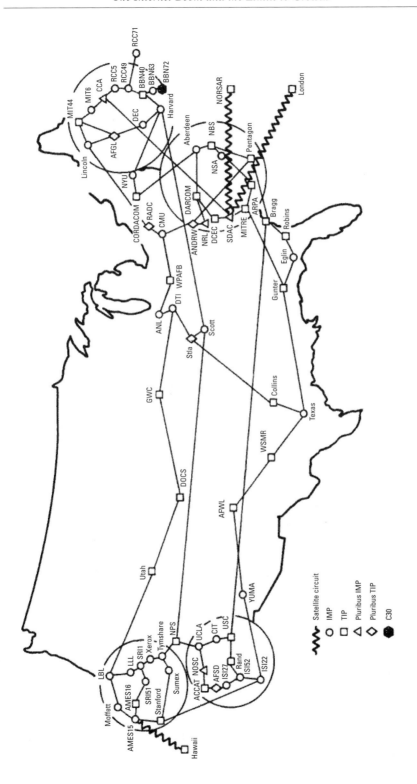

Figure 4.1 ARPAnet map from 1980 [3].

mainframes, the terminals were typically intelligent devices, block-mode terminals like the 3270 family, that performed local screen editing under tight control of the host. In the alternative computer universe then dominated by minicomputer vendors such as DEC and Hewlett-Packard Inc. (HP), terminals were typically "dumb" character-by-character asynchronous devices. AT&T's own Teletype Corp. had been a large supplier of terminals to minicomputer users in the 1960s and early 1970s, but its slow teleprinters (designed primarily for use as Telex machines) were being replaced by backward-compatible cathode-ray-tube terminals. In either case, the minicomputer terminal was more likely to be hard-wired to a small local machine. Big mainframes, on the other hand, were more likely to be located in a "glass house" computer center some distance away, so terminal-to-host data communications, or "teleprocessing" technology, was important. Minicomputer users were more concerned with host-to-host interconnection.

One very important network technology was the packet-switching protocol suite based on CCITT Recommendation X.25. This was developed in the mid-1970s as a way to reconcile competing packet-switched networks developed in several countries. (Like many CCITT Recommendations, as their standards were formally known, there was substantial wiggle room; each X.25 network had its own "profile" and actual devices attached to it required significant customization.) The design center was connecting asynchronous terminals to remote hosts, as an alternative to using modems on long distance telephone calls. The X.25 protocols were optimized for the slow, noisy analog telephone lines of the day, when a dial-up connection typically went no faster than 300 bps, and a very costly leased-line modem could only reach the blazing speed of 9,600 bps if the line was specially conditioned. Thus X.25 [4] had both hop-by-hop and edge-to-edge error correction and flow control, which the PTTs thought necessary. X.25 became ubiquitous across Europe by the early 1980s but never achieved more than niche status in the United States. New X.25 deployments largely halted by the mid-1980s. Many have since closed, but a few networks remained in place beyond the turn of the twenty-first century. They remained at least marginally useful for some niche applications such as credit card verification.

The American market for remote terminal connectivity was dominated by two carriers, Telenet and Tymnet, although other companies, such as AT&T and some of the Bell companies, also participated. Telenet was originally started by Bolt Beranek and Newman as a commercial spin-off of its ARPAnet technology, though its own technology took a very different path as X.25 and the ARPAnet diverged. By the 1980s, it was owned by GTE Corp., and it eventually became part of Sprint. Tymnet was started as the remote-access arm of computer timesharing company Tymshare. It outlasted its parent company, and eventually became part of British Telecom and later MCI. Both did provide optional X.25 interfaces; both faded into obscurity by the 1990s along with the terminal-to-host market.

IBM's original block-mode terminals used a protocol known as *bisync* (for "binary synchronous"). This was a simple Ack/Nak technique: Blocks were transmitted and the other side had to acknowledge each one before proceeding, or the block would be retransmitted. This limited throughput when transit delay was high, because only one block could be outstanding at a time for any given terminal. In the early 1970s, IBM saw satellites as the way of the future. It invested in Satellite Business Systems (SBS), hoping to bypass the Bell System monopoly with direct-to-the-business satellite service. The delays inherent in geosynchronous satellites were harmful to bisync. This gave them an impetus to develop a new protocol suite, called Systems Network Architecture (SNA). Although SNA eventually had some host-to-host connectivity, it was primarily designed for remote block-mode terminals. Large corporations built worldwide SNA networks, primarily to connect desktop terminals to large mainframes. It remained popular until terminals themselves were replaced by desktop computers in the 1990s.

Indeed the IBM mainframe world had its own sort-of-version of the Internet. Bitnet (whose name allegedly meant "because it's there") interconnected numerous universities beginning in 1981. Although not a packet-switched network, Bitnet used store-and-forward techniques said to be based on punch-card images. It was used for electronic mail and mailing lists. It grew to more than 500 organizations and 3,000 nodes (hosts) by 1989, but by 1994 most of its users had migrated to the TCP/IP-based Internet [5]. A closer analog to the Internet was Digital Equipment Corporation's DECnet. Begun in the 1970s as a way to interconnect the company's minicomputers, DECnet became a popular networking technology in the 1980s. It was very similar to TCP/IP; indeed, many of the same individuals worked on both. DECnet was not, however, designed for inter-company communications. Its most popular version, Phase IV, was limited to a 16-bit address field and 6-character host names. Within the DEC computing world, though, DECnet had many interesting capabilities; applications could often open files on any other computer on the same network just by prepending the host name and the characteristic double colon to the file name. Rather weak on terminal to host support, it eventually had Internet mail and file transfer gateways. DECnet's design center and function set were quite different from SNAs, but the two protocol suites were widely viewed as archrivals. DECnet has continued to be used in some applications, long after its creator was absorbed into Compaq Computer, itself later absorbed into Hewlett-Packard Inc.

The 1970s and early 1980s were thus a period of great diversity and progress in computer networking. That diversity would ironically enough turn into a monoculture by the 1990s, driven by the public Internet.

OSI, the Big Committee That Couldn't

Although DECnet and SNA were widely used, they had plenty of competition from other vendor-proprietary network technologies. One should perhaps make note of the pioneering work of Datapoint Corp., a Texas-based minicomputer company whose late-1970s desktop computers had the first commercially practical local area network, ARCnet, which included both its own coax-based physical media and the higher layers to enable data sharing. Xerox Corp. invented Ethernet, which evolved into the dominant family of LAN technologies even as Xerox itself drew little benefit from it. (DECnet Phase IV was itself designed in large part around Ethernet, of which Digital was a strong supporter.) Xerox developed its own protocol suite, XNS, which saw some usage in the early 1980s. Although largely forgotten, XNS was adapted by Novell Corp. into its own IPX/SPX protocol suite, which usually ran over Ethernet. Novell became the dominant vendor of LAN servers in the 1980s; it gradually migrated its servers to use TCP/IP only after the Internet was ubiquitous. Other also-ran protocols included Burroughs Corp's BNA and a host of LAN-based suites that took on Ethernet and Netware in the 1980s.

As computer companies developed their own technologies, many realized that their customers were, for the most part, not single-vendor shops. Although early computer networks were almost entirely single-owner affairs (the ARPAnet being Uncle Sam's private property, albeit with a few invited guests), vendors needed to make their computers interconnect. And so in 1978, the International Organization for Standardization (ISO) began an incredibly ambitious project. Open Systems Interconnection (OSI) promised to be all things to all people. And that was its downfall. It tried to be both SNA *and* DECnet, both X.25 *and* the ARPAnet, both meat *and* potatoes, and instead became a mish-mash of confusing options.

The most famous part of OSI is, no doubt, its Reference Model (the OSIRM). Defined early on, OSI's famous seven-layer model became *the* standard method of describing computer networks. Essentially every other protocol suite has been characterized in terms of the OSIRM. One might imagine, then, that the OSIRM had been well researched before it was adopted. But the truth is a bit more prosaic. The layered model was primarily created to divide the work among various subcommittees. Memorialized in the organizational structure of a huge international effort, the OSIRM was left intact long after it turned out to be, in part, unworkable!

Not that the OSIRM was entirely useless. It did a good job of characterizing packet-switched networks at the lower layers. Layer 1, the *physical* layer, transports bits. Layer 2, the *data link* layer, groups them into blocks called *frames* and provides for error detection. Together, these two layers generally operate on a hop-by-hop basis. Layer 3, the *network* layer, transports *packets* from one end of the network to the other and is the primary domain of *intermediate systems*, which today are usually called routers. Although the Internet Protocol itself is clearly a Layer 3 protocol, OSI's Layer 3 early on ran into an impasse in committee. One faction, largely interested in efficient terminal-to-host connectivity, wanted a *connection-oriented* Layer 3. It demanded, and got, a standard based on X.25. Another faction, more interested in host-to-host connectivity, wanted a *connectionless* Layer 3. It demanded, and got, one too. The ISO Connectionless Network Protocol (CLNP) was functionally very much like IP, which had preceded it, but with improvements such as a larger and more flexible address space. So OSI ended up with two parallel Layer 3 protocols to choose from both standards.

Layers 4–7 generally operated only within the host systems at either end of the connection. Layer 4, the *transport* layer, provided end-to-end integrity and flow control, running at the two ends of the connection. The OSI Transport Protocol ended up with five classes defined, one of which, called TP4, was useful over CLNP or X.25. The others were somewhat simpler but required the more elaborate Layer 3 services of X.25. TP4 was rather like a cleaned-up TCP.

The higher layers, however, were a disaster. Layer 5, the *session* layer, and Layer 6, the *presentation* layer, had uncertain functions that seemed to some to be a good idea in 1978 but which proved to be unworkable [6]. Layer 7, the *application* layer, included a set of protocols that were, as a rule, more elaborate than their Internet equivalents. By the late 1980s, it was clear that the functions originally posited for Layers 5 and 6 were really best left as options for the application layer. But many implementations had compounded the problem by trying to implement Layers 5 and 6 as discrete software, rather than within the application-layer code, which would probably have been much easier. It was not pretty. By the time workable implementations were reasonably commonplace, in the late 1980s, TCP/IP had already become the real open protocol for interconnecting systems. OSI was the big engine that couldn't.

TCP/IP Becomes the Standard

The ARPAnet was the government's own playground, but its protocols were not. In the early 1980s, the University of California at Berkeley undertook a major effort, fueled by cheap graduate-student labor, to create a form of Bell Labs' Unix operating system. The Berkeley Software Distributions (BSD) were an early example of what is now called *open source* software. Along the way, the TCP/IP protocol stack was implemented, and the Berkeley code became available to anyone wanting to use it. The price, under the BSD license, was to merely acknowledge the university's copyright.

Berkeley Unix became widespread in the mid-to-late 1980s. Sun Microsystems, for instance, got its start selling inexpensive hardware that ran a BSD-based Unix. So did many other workstation vendors, most of whom did not have the same staying power. Other operating systems also gained TCP/IP implementations, some ported from BSD. The Berkeley code was not always elegant or fast, but the price was right, and it did, in general, provide multivendor interconnection. Even the humble MS-DOS PC could run TCP/IP, although it was not until Windows 95 that Microsoft included it.

So corporate networks were developing around this free and surprisingly effective protocol suite. Numerous companies provided infrastructure support: Cisco Systems, for instance, was another mid-1980s start-up whose routers became very popular for corporate backbone networks. Its later good fortunes were based on its having embraced TCP/IP even while the smart money was on OSI. Digital Equipment Corp., in contrast, spent hundreds of millions of dollars on OSI. It was a network leader in 1985, yet by 1990 it was a rapidly declining also-ran, having been late to the TCP/IP party.

Digital was not alone in missing TCP/IP's rise. OSI was a huge project, getting huge press and academic backing. Even the Department of Defense, sponsor of the ARPAnet, had jumped on the OSI bandwagon. It had promulgated a rule, called GOSIP (for Government OSI Profile), that required computer systems vendors to have OSI support by 1990. (The United Kingdom had its own GOSIP.) Some people took this to mean that the government's TCP/IP networks would be replaced by OSI. But it did not work that way. Not only did GOSIP never really take hold, but it was only a vendor checklist item. Government procurements would have to have OSI available; GOSIP did not mean that anyone would actually use it. Large OSI networks were never built.

The Acceptable Use Policy

So by the late 1980s, there were all these corporate networks supporting TCP/IP, a protocol suite that was designed for internetworking. They were all dressed up with no place to go. Some companies and many universities had access to the Internet, of course. Starting in 1987 and with the help of then-Sen. Al Gore Jr. (D-Tenn.) [7], the National Science Foundation took over the funding of an Internet backbone for universities and research, the NSFnet. The ARPAnet itself was retired, whereas the Department of Defense maintained its own networks. NSFnet in turn had several regional "mid-level" networks tied together. Among them were New York's NYSERNET, New England's NEARNET, Southern California's CERFNET, and SURANET, collectively owned by several dozen Southern universities.

Taxpayer-funded networks were for government-approved applications. They had an Acceptable Use Policy (AUP). Commercial use was not permitted. A company could build its own TCP/IP network, and might even interconnect it to another company's, though it was not common. But if companies were lucky enough to get an attachment to the NSFnet, or to another network, such as CSnet, that had NSFnet connectivity, they certainly could not use it for purely commercial purposes. The concept of e-commerce was utterly anathema, and spamming was unheard of—it would have meant immediate disconnection from the network, if not worse. Indeed, there was a well-known story about a DEC salesman who e-mailed product announcements on the ARPAnet and was greeted with a reaction somewhat like a pork barbecue at a mosque.

The spam-free private-club nature of the Internet in those days was certainly pleasant, compared with today's cesspit of spam, worms, spyware, and script kiddies. But it left a large unsated demand. Internetworking was very attractive. But who would run the network? The Internet already existed. It had grown to have thousands of organizations in dozens of countries. A commercial network without connection to *the* Internet was seriously handicapped. By the early 1990s, the time had come for the Internet to be privatized and opened up.

Commercialization at Last

The phase-out of NSFnet funding and its replacement by private capital was not entirely smooth. NSFnet itself was operated by private contractors; it was nominally "privatized" in 1990. Advanced Networks and Solutions (ANS) was founded in 1990 [8]. It had three owners. One was the University of Michigan, whose MERIT subsidiary operated a regional network. Another was MCI, which brought bandwidth; the third was IBM, which brought computer equipment. In 1991, ANS asked the NSF for permission to provide commercial connectivity to the Internet. This would have provided funding to increase the network's speed; by that point the backbone was mostly T1 lines (1.5 Mbps). In that same year, the Commercial Internet Exchange (CIX) was founded; it provided a *peering point* where commercial TCP/IP network operators could interconnect without using the NSF backbone, and thus not be subject to the Acceptable Use Policy.

ANS thus planned to operate the federally funded NSFnet and its own ANSnet using the same facilities; it was really one network. ANS had a commercial unit, ANS CO+RE [9], which received permission to interconnect commercial users to the backbone, bypassing the AUP. Other operators wanted the same permission. ANS CO+RE's proposed special status did not go over well with the Internet community. By 1992, NSF wanted to phase out its own funding anyway. So the AUP was ended, and soon the mid-level networks became Internet service providers (ISPs). NYSERNET spun off a commercial entity, PSInet. BBN, which had operated several of the mid-levels on contract, purchased NEARNET, the Bay Area's BARRNET, and the South's SURANET; these became the foundation of BBN Planet, a backbone ISP that eventually became

Genuity (whose post-meltdown bankruptcy assets were picked up by Level 3 in 2002). All of these businesses seemed, in 1995, to hold great promise. Of course their long-term future was not nearly so bright.

Other companies jumped into the Internet game, of course. The single largest of the new commercial backbone ISPs ca. 1995 was probably UUNET. That company had begun in the 1980s providing any willing payer with electronic mail and newsgroup access using the UUCP protocol. UUCP was a message-switching program built into Unix; its name meant "Unix to Unix Copy." Non-Unix implementations existed, but UUCP was closely linked to its Bell Labs roots. Indeed it may have been one of AT&T's more profitable inventions: UUCP made use of dial-up modems to exchange files periodically, increasing the parent company's long distance revenues. The trouble with UUCP was that it was entirely ad-hoc, with only manual routing. A user sending mail to another user would have to specify the entire path of the message, separated by "bangs":

To: node1!someothernode!ihnp4!foobar!jones

UUNET offered a single hub for access, as well as a repository of files downloadable by dialing a 900 number. In the years leading up to Internet privatization, it became a major player in the UUCP-based network, known as Usenet. And as Usenet discussion-group (*news*) traffic migrated to TCP/IP, UUNET followed. In 1995, it was acquired by Metropolitan Fiber Systems (MFS); shortly thereafter, MFS was acquired by WorldCom.

Sprint also got into the business early, and for a time it had a natural advantage over BBN Planet, PSInet, UUNET, and other backbone competitors. Not only did it have experience in commercial data networking via the network that had begun as Telenet, but it also had plenty of its own bandwidth. It thus did not have to go outside and purchase bandwidth in an increasingly tight market. And in 1995, America Online acquired ANS, which it sold to WorldCom two years later.

Peering and Tiering

In some respects, the Internet resembles a large telecommunications network, but it developed very differently from the telecommunications industry. The rapid commercialization of the Internet created a very different competitive environment, a true free market in which there was no dominant player. ANS CO+RE had not been allowed special status, after all. Unlike the telecommunications industry, where regulation took the place of competition, the Internet business was not only subject to cutthroat competition, but it was deathly allergic to any kind of regulation. And how could it be regulated? Regulators take years to make decisions, yet the Internet was still like a child, growing and changing rapidly. Absent any dominance, there was no justification for economic regulation, be it telecom-style or antitrust.

As early as 1991, CIX members had agreed to exchange traffic with each other at no charge [10]. ISPs would be *peers* of each other, billing their subscribers and sending the traffic to the destination. Both ISPs would benefit by the greater connectivity. As the commercial Internet developed over the next several years, a somewhat more elaborate system evolved. A group of large backbone providers, who tacitly agreed among themselves that they formed the core of the Internet, agreed to be peers with each other, but they sold their services to smaller "downstream" ISPs. A tiered network developed. Tier 1 providers had national backbones with direct interconnection to each other, had connections to the major traffic exchange sites that developed, and agreed to exchange traffic among each other both for their own direct customers and on behalf of the lower-tier ISPs that used them for "upstream" service [11]. Tier 2 providers also had national networks that agreed to peer with each other and with the Tier 1s, though they may have had to pay for "transit" when relaying others' traffic across Tier 1 providers [12]. About two dozen providers probably warranted being viewed as Tier 2. Most regional and retail ISPs purchased upstream service from a Tier 1 or 2 provider.

The price of upstream service was, like the price of commercial access itself, usually based on capacity, not on measured traffic. An ISP would have a price per megabit or for a given type of interface, such as DS1 or 10 Mbps Ethernet. Since IP was connectionless, there were no calls that could be easily metered. Raw packet or byte count could be, but a packet to an in-house server would be indistinguishable from one going around the world. Geographic rate averaging ("postalized" pricing) was as much a practical necessity as a conscious decision.

An Industry Structure Develops

The earliest ISPs were, for the most part, in the connectivity business. PSInet, ANS, and BBN were selling businesses and institutions, such as universities, access to the developing global Internet. Their customers had their own computers—their own servers—and thus did not need much more than a big pipe. This turned into a commodity business. Although the performance of different networks was not identical, the basic service was quite straightforward.

But consumers were a different business altogether. Even before 1993, several million American consumers had modems and subscribed to on-line service providers such as Compuserve, GEnie, Prodigy, AOL, Delphi, and The Source. These services provided electronic mail, file transfer, chat, on-line forums, and other information services. By 1994, a consumer ISP business was starting to develop. These companies did not need the complex information infrastructures of the earlier on-line providers, because they offered the Internet's wealth of information instead. The invention of Mosaic, the first graphical World Wide Web browser, in 1993 essentially made the earlier information services obsolete. Retail ISPs instead offered dial-up modem access, mail servers, and, in most cases, Web servers for subscribers to put up their own pages.

Most retail ISPs, unlike the national on-line service providers, initially owned their own modems. But operating modems became a specialty business of its own. AOL, for instance, was one of the largest of the on-line service providers, with three million subscribers, at the beginning of 1995. At that time it was beginning to phase out its use of Telenet and Tymnet's X.25 networks and install a new "AOLnet." The latter was not internally operated by AOL; it was outsourced to Sprint, BBN, and ANS (the latter, however, owned for a time by AOL). It was originally going to use UUNET, but when Microsoft announced its own competing service, MSN, UUNET took an investment from Microsoft and became MSN's key modem provider; its role in AOLnet went to BBN instead. Various other companies also set up "rent-a-modem" services, enabling retail ISPs to have a larger local-calling footprint than they could afford to provision on their own.

Thus the ISP industry developed three major subindustries (see Figure 4.2):

- *Access* ISPs interfaced with local telephone companies, providing modems and later digital subscriber line (DSL) services in bulk.
- *Vertical* ISPs operated servers and dealt directly with retail customers. Vertical ISPs included both consumer-oriented retailers and commercial-oriented operators, such as Web hosting companies.
- *Backbone* ISPs provided connectivity between each other and to vertical ISPs.

Some providers, of course, operated in more than one category. But specialization became the norm; even companies operating in more than one subindustry tended to have a natural separation between the lines of business. Backbone providers sold directly to large businesses and to downstream vertical ISPs, a business-to-business sales model built on large sales departments and small billing departments. Vertical providers selling to consumers had a mass-market sales model, typically billing via credit card. Few companies could make both sales models work together, so even those who began trying to be full-service providers often abandoned some markets.

Web hosting was a form of vertical ISP business that evolved naturally out of backbone provision. Before the World Wide Web became such a dominant application, companies with Internet connections generally maintained all of

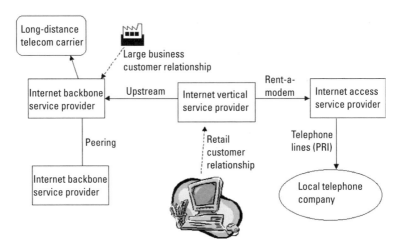

Figure 4.2   Internet service provider industry model. Arrows indicate cash flow.

their own computers on site. And the first Web servers were generally at their companies' sites, just another machine in the data center. But as Web bandwidth grew, the cost of local bandwidth, from the ISP backbone to the customer, became greater. Because public Web servers had little community of interest with their owners' own internal networks, it made sense to bring them to the backbone, rather than bring the backbone to them. This led to the growth of the Web hosting industry, led by the backbone ISPs. So, for instance, in 1995, BBN Planet introduced its Web Advantage service, and located several racks of Web servers in a large computer room that had once housed the parent company's DEC-10 mainframes. As the service grew, more of the company's internal computers were pushed aside to make room for Web servers, and more "Web farms" were built in California and elsewhere, at sites where the company's backbone routers were located. Other backbone ISPs did the same, of course. This then led to a number of start-ups of companies whose sole business was hosting other companies' servers. Perhaps the best known was Exodus Communications, founded in 1994, which grew to 48 data centers by the time it went bankrupt in 2001 [13]. Exodus did not, however, own its own backbone transmission facilities, which may have contributed to its inability to turn a profit.

Internet Traffic Explodes as the Public Joins

The amount of bandwidth on the Internet had been growing at a rapid pace for many years before the NSFnet backbone was privatized. The ARPAnet's backbone links, into the early 1980s, were largely made up of 50,000 bps circuits. That was no small feat to achieve, because an analog voice-grade leased circuit at the time could, in general, only support a 9,600 bps modem. The backbone was built using *group channel* modems, Bell type 303, each of which consumed the bandwidth of *12* voice channels on the AT&T Long Lines network. By the time of the NSFnet, 1.544 Mbps "T1" circuits were readily available, and by the time of privatization, the busiest backbone routes were starting to migrate to the DS3 rate of 44.7 Mbps, roughly a thousand times faster than the old ARPAnet backbone. That was only the beginning.

Internet backbone bandwidth requirements were growing rapidly for two reasons. One was the increased bandwidth used by the average subscriber, as new applications, lower-cost and higher-speed links, and faster computers facilitated more intensive use. The World Wide Web only began to catch on after the introduction of Mosaic in 1993, yet by 1996 it was widespread, making the Internet far more attractive to consumers. Electronic mail traffic grew as more people became reachable. And the opening of the network to the public created the scourge of spam, as unscrupulous marketers began to harvest e-mail addresses from any possible source, and sell them to each other.

Far more importantly, the number of users was increasing rapidly. Internet service providers were popping up everywhere; although counts were not definitive, *Boardwatch Magazine* was estimating the number of ISPs in the United States and Canada, in August 1997, as 4,133 [14], up from only around 90 in 1993 [15]. The number rose much more slowly over the next couple of years, stabilizing as some small providers consolidated. The older on-line service providers either faded away or became ISPs. Some of the largest bulletin board operators also became ISPs; the "BBS" business itself rapidly faded.

Data Traffic Finally Tops Voice

It is hard to tell exactly when the telecommunications industry carried more data than voice traffic, but it probably occurred during the 1995–1997 explosion in Internet traffic. To be sure, it is hard to precisely quantify the actual amount of data carried across the Internet. The TCP/IP protocol suite is rather inefficient, so many links are run at a fairly low percentage of utilization. Bandwidth is often purchased by ISPs in bulk, ahead of demand, in part because telecommunications providers make bulk bandwidth available in rather coarse multiples (most often 1.544 Mbps, 44.7 Mbps, 155 Mbps, 622 Mbps, and 2,488 Mbps). So bandwidth provided in bulk to an ISP is not necessarily being fully utilized. But it creates demand for the telecommunications providers. And it was the data providers that were sucking up the available bandwidth, not voice-oriented long distance networks.

By 1996, the fiber-optic and SONET networks, installed during the preceding decade by AT&T, Sprint, MCI, and WorldCom's various predecessors, suddenly found themselves running low on capacity. The long-term growth rate for voice traffic had been a few percent a year, giving plenty of time for planning. Competition in the long distance business had led to lower prices and increasing demand but at manageable rates. Interstate long distance calling rose 158% between 1985 and 1995, but the fastest growth was in the late 1980s [16]. Data traffic was growing much more rapidly. ISPs were suddenly having trouble getting DS3 circuits to add to their backbones.

To be sure, fiber optics themselves have *huge* potential capacity. But upgrading the capacity on a given *strand* requires new opto-electronic equipment. Lighting up new strands of fiber often requires new repeaters and other gear to be installed along the path. Wavelength-division multiplexing (WDM) had not become commonplace by 1996, so fiber strands were running out of capacity relatively quickly.

Basic economics, going back to Adam Smith, indicates what is likely to happen when a commodity—for that is all that raw bandwidth is—is in short supply. Prices may rise, and new suppliers are tempted to enter the market. But the telecommunications business, especially the long-haul backbone bandwidth market, is not like the ones that Smith dealt with in eighteenth-century Britain. It is a large-dollar long-term investment. It only makes sense if there is a high probability that supply will not severely exceed demand, which would lead to a precipitous fall in price. Demand forecasting is thus critical.

Short-Term Versus Long-Term Trends

Many people believed, during the boom, that Internet traffic was doubling roughly every 100 days. This fantasy was based on statements made by WorldCom in the 1997 time frame. It was widely repeated, becoming the basis of many market research firms' forecasts of demand for all things Internet. It led financiers to put up *trillions* of dollars in capital. After all, demand would soon catch up with whatever supply could be built. Right? Wrong.

WorldCom's estimate was no more precise than its corporate accounting turned out to be. David Faber of CNBC traced it to Tom Stluka, an employee of WorldCom's UUNET subsidiary. Stluka "created a best-case scenario for the Internet's growth," reported Faber [17]. "Stluka's model suggested that in the best of all possible worlds Internet traffic would double every 100 days." Stluka was not reporting on growth but speculating on sales forecasts. But it got a life of its own, and WorldCom's 1998 annual report stated—based on Stluka's forecast, by then much retold—that the Internet was growing at 1,000% a year. WorldCom later claimed that there was some truth to the number: Its backbone network's bandwidth capacity was indeed growing extremely rapidly during a brief period of time, when UUNET was rapidly building out its network and catching up with demand [18]. But even that excuse does not quite ring true.

Even if actual Internet traffic *had* doubled during some 100-day period in the mid-1990s, short-term growth does not extrapolate into a long-term trend. Growth rates in new technologies frequently follow an "S"-like curve. The growth rate turns suddenly steep as early adopters give way to the mainstream, then levels off further as the market saturates.

Andrew Odlyzko's study, *Internet Traffic Growth: Sources and Implications*, reported that traffic on U.S. Internet backbones rose from 16.3 Tbytes per month in 1994 to 1,500 Tbytes per month in 1996. This hundredfold growth reflects the opening up of the Internet to the broad public, with huge pent-up demand fueling a one-time burst of growth. He estimated 5,000–8,000 Tbytes per month in 1998. This is more like 100% per year than 1,000%. Traffic doubling annually is certainly impressive, but it would not have been enough to fuel all of the investments made during the boom. The doubling-every-hundred-days estimate was instead repeated as if it were gospel truth, a permanent rule of nature, becoming a favorite of the Wall Street "analysts" who blessed every cockamamie proposal to come along.

WorldCom Set a Suspicious Pace

It is not entirely a coincidence that so much of the meltdown came about because of investments built on a WorldCom lie. WorldCom itself came to embody the era, a company that seemingly "got" the Internet. After it had acquired MFS, a pioneering competitive local exchange carrier (CLEC), and with it UUNET, the leading ISP, WorldCom had put together the most integrated telecommunications company since the breakup of the Bell System. It owned a long distance backbone, an ISP, local dial tone, and fiber-optic local loops of its own.

How did WorldCom get so big so fast? The company had only been founded in 1983, by a group of businessmen largely funded by Bernard Ebbers, a Canadian who had become a basketball coach and then motel owner in Mississippi. Hardly the type of background one expects from a technology titan, Ebbers' company was originally called Long Distance Discount Service. Over the next decade it had a ravenous appetite for acquisitions. It rolled up dozens of small long distance companies, including Advanced Telecom Corp. (ATC), American Network, Claydesta Digital, Microtel, Mid-American, and NTS, among others [19]. It acquired the long distance operations no longer wanted by corporate giants Metromedia and ITT. By the end of 1994, it had a 1.3% share of the long distance business [20]. Ebbers' style of acquisition was aggressive. He was not afraid of a "Pac-Man" acquisition, in which his company swallowed up a larger one whole, remaining the controlling party. That required a high stock price. He could then buy companies for stock, rather than cash. His stock in trade was his stock, in trade. The high stock price was not based so much on earnings as on a high price-earnings multiple, based on investors' expectation of rapid future growth. Companies with a high multiple live in a world of eat or get eaten. This was a good example of the management style described by Rolf Wild in his 1978 book *Management by Compulsion: The Corporate Urge to Grow* [21]. Profit and stability are, in a world of hostile acquisitions, secondary to a rapid growth rate, which is what produces a high multiple. A company that has a high multiple can stage friendly or, if necessary, hostile takeovers of competitors with lower multiples. Ebbers did not want to lose control; he wanted to be in control.

In 1995, LDDS acquired Wiltel, a major provider of bulk bandwidth to other carriers. This made it the fourth-largest player in the business market, right behind Sprint. Wiltel's fiber-optic network had been built along the natural gas transmission rights-of-way of its parent, Williams Companies. Its Points of Presence were, not coincidentally, often at the same locations as MCI's; Wiltel was obviously selling bulk bandwidth to the number two provider. LDDS changed its name to LDDS WorldCom, dropping the LDDS name not long afterwards. (The WorldCom name itself was apparently inherited from IDB WorldCom, an international satellite network provider that it had acquired less than a year earlier.) With Wiltel's network under its belt, WorldCom was suddenly a player to be taken seriously, a major force in the marketplace. It did not have a compelling presence in the fairly stagnant retail market, but it was a major supplier of bulk bandwidth to the burgeoning ISP market. Ebbers had jumped from a slow-moving train onto a fast one indeed.

WorldCom's eventual collapse was, perhaps, inevitable, given the nature of its business. But that was not the telecommunications business, or for that matter the Internet business. WorldCom's business was acquiring companies. It was not good at integrating them into a whole; order processing, billing, and service were all quite disjoint. But there was no time to fix that; the company's compulsion was growth by acquisition.

Once WorldCom was number four, acquisition targets became scarcer. MCI had, for several years, been partnering with British Telecom (BT). In 1997, BT made a friendly offer to acquire the 80% of MCI that it did not already own for $21 billion, though the offer was soon lowered to $17 billion [22]. WorldCom replied with an all-stock bid for all of MCI worth, at the time of the offer, about $30 billion. GTE Corp., the largest of the non-Bell local telephone conglomerates and by then the owner of BBN Corp. and its Internet backbone business, counter bid $28 billion in cash. The bidding war ended with WorldCom's winning offer valued at $37 billion.

MCI had, by the end of 1997, become a key backbone ISP, not quite as big as UUNET but a major force. To assuage antitrust concerns and guarantee that no company would become a *dominant* player in the Internet, MCI's Internet business was sold off in the merger to Cable and Wireless, the old British international carrier. But even before the MCI merger was closed, WorldCom acquired Brooks Fiber, a major competitive local exchange carrier whose footprint had little overlap with earlier CLEC acquisition MFS. Of course MCI also had a large CLEC; within three years of the Telecom Act's authorization of the CLEC industry, WorldCom was by far the largest CLEC.

Where could WorldCom go next? Having bought the number two company in the long distance business and already being the largest backbone ISP, the growth rate was bound to decline. The die was already cast for WorldCom's decline; the company's accounting became more and more creative to maintain appearances. Its last gasp as an acquirer was an attempt to buy Sprint Corp. Sprint was the number three long distance provider and, like UUNET and now Cable and Wireless, one of the top Tier 1 backbone ISPs. But it was 1999, the peak year of the boom, and WorldCom's valuation was still high. Sprint was not like WorldCom, though. It had started out as United Telecommunications, a rural local exchange carrier, and had acquired the long distance company from GTE several years earlier. Its major speculative investment was its wireless network, Sprint PCS. It was not a profitable operation, but WorldCom's wireless holdings were very limited (it resold other carriers' cellular service) and wireless was a hole in its product line.

The Sprint deal fell through in 2000, largely because of opposition from Europe. And WorldCom's stock began its inevitable fall. It reached its peak market capitalization, $120 billion, in 1999, but its rapid fall continued until it filed Chapter 11 bankruptcy in 2002, after its fraudulent accounting had been exposed. The specific sin that brought WorldCom down was that it inflated earnings by booking expenses—in particular, access payments to local exchange carriers—as capital, therefore improving short-term profit margins. The post-bankruptcy company emerged under new management, renamed MCI.

ISP Pricing Creates Permanent Losses

With its scale and with WorldCom's bandwidth in-house, UUNET may well have had the lowest cost of operation of the original Tier 1 ISPs. AT&T was a Johnny-come-lately to the business; it started dabbling in the backbone business in 1995 by reselling BBN Planet's service, starting its own the following year. MCI had sold its interest in ANS and was starting to build its own new ISP backbone in 1995, catching up with Sprint. These companies could afford to quickly add bandwidth and lower their prices at the same time. But other ISPs were not so lucky. BBN Planet, PSInet, AOL's ANS, and other ISPs paid others for their bandwidth.

Their other expenses were high as well. Backbone ISPs had high selling costs. Large retail ISPs such as AOL and Earthlink relied on advertising and, in AOL's case, vast amounts of direct mail. That was costly enough. But the backbone providers made considerable use of the direct sales model, with salesmen making calls on potential customers. This required an infrastructure of regional sales offices and sales support. And the Internet business was *not* cut-and-dried. Although managers tried to make it a "repeatable" process, customers' needs were changing as fast as the network was growing. (And even if it was not doubling every 100 days, doubling once a year was plenty fast.)

The nature of the Internet backbone service itself may have contributed to the ISPs' financial woes. IP packet relaying, the core business of backbone providers, is usually referred to as a "best effort" service. This is a euphemism for "no particular effort," based on the concept that the provider will make its best effort, but not offer a guarantee, to deliver the packets. When there is only one class of service, best effort is also worst effort. The different backbone providers had to exchange traffic with one another, of course, but since peering was "free," with no financial incentive to provide other ISPs with a particular rate of delivery, or any defined quality of service. As Gordon Cook described it, "Best effort has no value beyond that of an ever-deflating commodity" [23]. Backbone providers were selling commodity-grade service. They could provide premium service *within* their own networks, but even that can be costly, and lacked associated revenue other than preservation of market share.

The old-line telecommunications business had a high expense structure too, but it had a simple solution: It set prices high enough, on average, to cover costs and make a decent return on its investment. Monopolies helped, of course. The ISPs, however, never had monopolies. They set prices at a level far below break-even. This was justified by competitive necessity: If one ISP raised its rates, then it would lose customers to one of the several others whose rates were at the lower, albeit probably unprofitable, level. At least one head of a Tier 1 ISP, however, explained his strategy to his employees [24] more openly. The idea was to wait until a shakeout happened, when its competitors ran out of money and left the businesses. Then the survivors could raise prices.

This was, of course, an impossible strategy. Nor did it work for that CEO. His company was, according to its forecasts, about two years away from break-even, according to its cost and revenue curves. And indeed break-even remained at least that far away until it went bankrupt some years later. But a wait-for-a-shakeout strategy could have failed anyway. Bankrupt companies did not necessarily shut their doors. They often shed their debt and continued to operate. Without debt, their costs were lower, and they were even stronger competitors. That indeed was the pin that pricked the bubble.

Investors Subsidized Prices

A company can only continue to sell below cost for as long as it has the capital to do so. But in the boom era, capital seemed to be the easiest thing in the world to come by. So the business model that most publicly traded Internet companies followed in the late 1990s was fairly simple: Sell stock at inflated prices and use the money *in lieu of* adequate revenues. The real product was stock; customers were merely icing on the cake.

This applied to both the ISP business and to the businesses that became their flagship customers, the so-called *e-commerce* companies, or "dotcoms." The stock value of publicly traded ISPs and Internet-related companies rose, along with that of many new telecom companies. Initial public offers were typically oversubscribed, and if the company ran low on cash, it could, during the boom, typically go back to the well for another round of equity. Privately traded companies also found equity easy to come by, as venture capitalists poured cash into anything even remotely related to the Internet. The goal, of course, was to cash out, usually by going public. So long as an IPO could bring in huge returns, venture money would be plentiful.

Take, for instance, Teligent, whose primary business was operating a CLEC using wireless connections to large buildings, which also operated an ISP, and which was headed by a well-reputed former AT&T executive, Alex Mandl. Its 1997 IPO raised $118 million, while Nippon Telephone threw in another $100 million of equity. Microsoft and Hicks Muse put in $400 million in convertible preferred stock, while more than a billion dollars of debt financing was obtained. That would certainly have been enough cash to build a sustainable business, if the company had a sustainable business model. By 1999, annual revenues were only $31 million, and losses were growing [25]. Its subsequent bankruptcy was not a surprise. But some insiders explained it differently. The earliest buildings on the network were beginning to return a profit. Had Teligent expanded more slowly, or had it gone out for an additional round of financing before the window closed, then the company might have been able to hold on until break-even. But slow expansion was not the order of the day. Everyone acted as if more investor money would always be available whenever needed [26]. When it dried up, the game was over, for Teligent and many others.

Dotcoms Create a Demand Bubble

Stock prices did not rise during the boom years because company fundamentals were improving. Internet-related stocks were rising in large part *because* they were rising. The stocks had momentum. Investors bought in because they saw rising prices, and assumed that the price would continue to rise.

This was, of course, a classic "bubble" situation, one that has happened many times. One of the oldest such cases was "tulipmania," when the price of tulip bulbs in Holland briefly rose to astronomical prices. This was documented in Charles Mackay's 1841 classic *Extraordinary Popular Delusions and the Madness of Crowds*. One bulb was bartered for goods worth, today, about $35,000 [27]! Of course, many skeptics viewed the Internet bubble as analogous to the tulip craze of 1635. Only the tulip first grew in western Europe in 1559, and it was decades before the price took off. The 1990s bubble happened only a few years after the Internet was privatized: "Internet time" moved so much faster.

The Internet stock bubble was quite broad. Internet service providers themselves, of course, were beneficiaries. These were all, almost by definition, rather young companies—although established companies like Sprint had ISP subsidiaries, the excitement was hottest for initial public offerings of "new economy" companies built around the Internet. "New economy" was an interesting euphemism. Promoters saw it as a bright future of an Internet-driven economy, where old-fashioned "bricks and mortar" businesses would fall to new competitors. But a more realistic view was that it meant an economy where suckers could be parted from their money without any of the traditional fundamentals that normally make a stock attractive.

The story of the dotcom companies is a long and, in some respects, a sordid one. It is not the primary focus of this account; the dotcoms are primarily important here because they generated demand for Internet and telecom services and equipment. As the dotcoms grew, so did their suppliers; as the dotcoms fell, so did the fortunes of companies that had invested heavily to support them. As with most new industries, of course, the dotcoms were not all failures. Some, like eBay, proved to be long-term profit generators, whereas others, like Amazon, appear to be survivors, if not terribly profitable. But the failures were spectacular. Companies founded by recent graduates in 1997 were worth tens of millions of dollars by 1999, resulting in the "dotcom millionaire" phenomenon. Most were worth zero by 2001.

The dotcoms got that nickname because so many of them put ".com" into their corporate names. This was itself turning the Internet's previous norms upside-down. The Domain Name System [28] (DNS) created a hierarchical network database to identify Internet nodes. The ARPAnet had previously used a flat name space, wherein every host computer maintained a HOSTS table giving the numeric address for every named node on the network. Computer names were unqualified, such as DECVAX and RUTGERS. DNS created the .com space for commercial ventures, alongside other top-level domains such as .edu for higher education, .gov for the U.S. government, .org for noncommercial organizations, and .mil for the U.S. Department of Defense. Other top-level domains were later added, including two-letter domains based on ISO country codes, but .com was by far the most popular address space, and the only real home for an on-line merchant. Although older companies typically selected their DNS domains by putting .com at the end of their name (e.g., sun.com and microsoft.com), Internet-oriented start-ups often put ".com" into their corporate names. Amazon Books had renamed itself Amazon.com Inc. after diversifying its product line. Other companies created their corporate identity by purchasing an easily remembered domain name from a speculator that had grabbed it earlier. This provided names to famous flameouts such as Pets.com and Toys.com.

Many observers mark the beginning of the bubble with the IPO of Netscape Communications in 1995. Netscape, still basically a start-up, was founded in 1994 by Mark Andreesen, who a year earlier, as a graduate student in Illinois, had written Mosaic, the first graphical Web browser. Netscape's main product was its browser software, which it gave away for free. The company had been founded as Mosaic Communications Inc., but the University of Illinois reminded it that the university owned the name Mosaic. Andreesen's new browser's nickname was Mozilla; early release notes said it was spelled Netscape but pronounced "Mozilla." [29]. In 1995, Web browsing was rather new, and Netscape's new browser was a market share leader. Netscape did, however, try to sell some later versions of the browser—an effort that met with little success—and it did have some for-pay products, such as a Web server. Its products were potentially attractive for corporate "Intranets," as well as public Internet applications, and some people viewed the browser as a new software-development platform, potentially as important as a computer's operating system. Still, a bet on Netscape's stock was at best highly speculative. The company expected to raise about $60 million to $70 million by going public, but at the end of its first day of trading, its market capitalization was an astounding $2.3 billion. This was an open invitation for other companies to jump in and capitalize on Internet tulipmania.

Netscape itself was not a long-term success story. The company's browser market share declined precipitously after Microsoft decided it wanted control of the browser. The software giant, ironically enough, began by licensing Mosaic itself from Spyglass Inc., which held the licensing rights from the University of Illinois' National Center for Supercomputer Applications, where Andreesen had worked as a student. The first releases of Microsoft Internet Explorer were not terribly impressive, but the product rapidly became competitive. Microsoft Windows 95 did not originally include a browser, but by 1997, Internet Explorer was included with new copies, and Windows 98 was designed to tightly integrate it into many aspects of the system. With every new Windows system including a perfectly serviceable browser, most users saw no need to download Netscape Navigator, even though it was arguably superior and available for free. Finally, in late 1998, America Online acquired Netscape for $4.2 billion. This was well below its peak, but still inexplicable in terms of fundamentals. AOL did not even use Netscape's software for its own service, which at the time was growing rapidly. Netscape's main attraction may have been its home page: The browser, as installed, defaulted to Netscape's own home page, which carried advertising, and thus presented a modest revenue stream.

The advertising model was the primary attraction behind another leading Internet stock, Yahoo. Founded as an Internet search company, Yahoo developed a number of services behind its "portal," many carrying advertising, and some, such as an electronic commerce shopping cart service for small merchants, carrying real user fees. It actually made a profit. But its market cap during the bubble was ridiculous. *Computerworld* noted in 1999, "Santa Clara, Calif.-based Web portal Yahoo Inc., for example, is worth more on paper ($59.5 billion as of late November) than Detroit-based General Motors Corp. ($47.6 billion), even though GM has more than 100 times the net income ($2.9 billion vs. $25.6 million)" [30]. Internet-related service companies also fared well for a short time. Two companies, Whittman-Hart and USWeb/CKS, merged in 1999 to form a company called marchFIRST. Whittman-Hart was the older, dating back to 1984; it had, however, discovered the Web and jumped on the bandwagon. The market capitalization of marchFIRST rose to $14 billion as the company spent money prodigiously, opening offices, furnishing them lavishly, and hiring young Web designers *in anticipation of* business that never arrived. This too was characteristic behavior of the era. When it lost $6.8 billion on revenues of $213 million in 2000, the only question was whether shareholders' money was vanishing faster than its employees' jobs. It failed shortly thereafter; its market capitalization in 2002 was approximately zero.

AOL itself was unusual in one respect. When its market capitalization was still high, in January 2000, it acquired an old-line business, Time Warner Inc., creating AOL Time Warner. Its market capitalization was twice Time Warner's, but its shareholders accepted 55% of the combined company's equity. The marriage was a rocky one. By 2003, AOL's core business of dial-up Internet access was in severe decline, the Time Warner part of the company was clearly in control again, and "AOL" was dropped from the corporate name. Had AOL remained a stand-alone company, no doubt its market capitalization would have been a small fraction of Time Warner's, whose own diversified media business was doing better. But because it had *monetized* its stock by merging with Time Warner, AOL's shareholders were left with far more than most other dotcom investors. Any number of dotcoms could, at their peak, have monetized their market capitalization by purchasing large, profitable old-line corporations. Hardly any did.

Carrier Hotels Created Too Much Room at the Inn

The first corporate Web servers were, most likely, located in their company's data centers. As the Web caught on in 1993–1994, many companies rushed to put up Web pages, but total traffic was still low, and it was possible to just use the company's own ISP connection to do so. However, that approach quickly grew tired. The biggest cost of an Internet connection was often the telephone company connection from the ISP to the corporate site. That was, after all, usually a monopoly, and prices of these circuits were set, like long distance service, at "contributory" levels. As demand grew, it became impractical to raise the bandwidth of this costly link just to present information to the outside world. So companies started collocating their Web servers at their ISPs' locations.

As previously noted, ISPs often had Web hosting services, taking advantage of their own backbone connections. Some IXCs and ISPs also rented out collocation racks as a business. They did not necessarily operate the servers, but they did provide secure, air-conditioned facilities where customers could get ample backbone bandwidth for their collocated servers. Wiltel, for instance, began hosting AOLnet modem racks shortly before being acquired by World-Com; it also provided the backbone connections to AOL's Virginia headquarters. By the time AOL's modem traffic peaked, WorldCom had hundreds of racks of modems in dozens of facilities. Many new dotcom companies did not even have large offices or computer rooms of their own; the bulk of their bandwidth never came closer than their ISP's collocation room. As the dotcom boom grew, space grew tight and rates rose. Racks were renting for $1,000 per month, sometimes more, and that did not include any bandwidth. Still, the collocation rooms filled up. Customers that needed more than a rack or two might want a room of their own; a 100-square-foot room in an ISP's building might go for $10,000 per month. And that was typically the smallest size room.

Soon, other companies simply opened up physical facilities in which others could collocate their servers. They would arrange for the telecom carriers and backbone ISPs to bring in fiber-optic connections and rent space by the rack. ISPs and telecom carriers were happy to show up because these buildings were full of potential customers. Renting a 23-inch space for $1,000 per month was certainly more than landlords could get for office space, even in the boom years' overheated market. Thus, Adam Smith's prediction came true again: Lack of supply caused an increase in price, which led to an increase in supply as more "carrier hotels" opened.

By early 2000, speculative announcements of new carrier hotels were well out of hand. In the Boston market, existing ISP-based facilities at WorldCom, Level 3, and Network Plus had high occupancy rates. An independent carrier hosting company called CO Space was doing well too; CLECs such as Global NAPs also had plenty of tenants. Of course many of their tenants were dotcoms with shaky business models that were not long for the world, but the real estate industry was not prepared to see that. Investors (to use that word in its loosest sense) were clamoring to enter the business. So a number of new facilities were announced. An old Raytheon factory in Waltham, Mass., the former Wonder Bread bakery in Natick, Mass., the upper floors of the former Jordan Marsh department store in downtown Boston, and an old warehouse in Allston, Mass., were all on the developers' agendas. None opened. The Allston building, prominently located along the Massachusetts Turnpike, was pretentiously named Boston Internet City. The developer got as far as removing the building's exterior walls by the time the carrier hotel business collapsed. The building sat for years sheathed in cloth before being redeveloped for the Next Big Thing, biotechnology. But that's another tale.

The Bubble Bursts in Equipment Manufacturers' Faces

The boom years were good to manufacturers of both telecom and Internet gear. This stands to follow; capital gear is highly cyclical, and the 1990s were both a strong business cycle and a focused one.

IP-network equipment providers had, before 1993, sold most of their gear to enterprise networks, a steady if unglamorous business. Some sales were made, of course, to the NSFnet members and other public providers, but absent public access to the Internet, that channel was constrained. The creation of an ISP sector practically overnight caused a surge in demand. Large routers were needed for backbone nodes and regional points of presence; small "edge" routers were deployed at customer sites. Going into the mid-1990s, the "big four" companies dominated the field: Cisco, Bay Networks (formerly Wellfleet), Cabletron, and 3Com were all successful in selling routers to enterprise customers. The public Internet boom primarily benefited Cisco, which pulled away from the competition. By early 2000, Cisco's market capitalization was the highest of any publicly traded company in the world. Even more interesting, Cisco's growth was largely organic, as it added products and grew market share in a fast-growing market, though to be sure it did acquire numerous smaller companies for their technology. Also-ran Bay Networks had merged with Northern Telecom, a major manufacturer of telephone equipment, becoming Nortel Networks; it never achieved Cisco's boom-era valuation but did share in the crash to come.

Another giant, Lucent Technologies, had only been spun off from AT&T as the boom was getting underway. It was to some extent a beneficiary of both the Telecom Act and the Internet boom. The Internet generated demand for second telephone lines to be used for modems. Lucent and Nortel, the two dominant central office switch vendors, busily filled orders from the incumbent telephone companies as the last of the analog switches were phased out [31] and the existing digital switches were expanded. At the same time, the ISPs themselves needed dial-in capacity, which was mostly met with T1 interfaces on the switches, typically configured as ISDN primary rate interfaces. So the ILECs placed orders for T1 ports; given the long lead times that they worked on, backlogs were reportedly over a year long. This led manufacturers, such as Lucent and Nortel, to increase capacity, which of course worsened the oversupply when the boom ended.

Surplus Gear Met Demand

A final feature of the meltdown was the mass of unsold equipment left over from the boom years. Surpluses continued to build up as CLECs, ISP, dotcoms, cable overbuilders, and other companies failed. Routers and servers became available for pennies on the dollar. Most of this hardware was unused, new in the box; slightly used equipment was even cheaper.

Of course one company's bad news is another's good news. Survivors and start-ups could build "eBay-powered networks" for a fraction of 1999's costs. But for the manufacturing sector, it was adding insult to injury. This was severe for router leader Cisco Systems, but it had the resources to weather the storm, albeit slightly slimmed down. The company reportedly spent billions of dollars purchasing back inventory in order to crush it, or at least to limit its impact on the surplus-gear market. This apparently helped at the high end of the market, where a relatively small number of ISPs and telecom carriers were the potential buyers and sellers, but the midrange remained saturated with available products.

The telephone switch business did not face the same surplus problem but for a different reason. Telephone switches are subject to strict regulatory requirements, which lead to regular software updates. Lucent and Nortel licensed their switch software on a nontransferable basis. Most of the cost of a new 5ESS was software; a used 5ESS would require the same license fees as a used one. Because new switches from vendors such as Sonus and Taqua could be had for a lower cost than relicensing alone, the market price for used legacy telephone switches fell to approximately zero, and few were sold. A brisk trade remained in specialty gear such as Excel switches, which often depended on third-party software, and in transmission gear. The price of an M13 multiplexor, the generic device breaking a DS3 circuit down into 28 DS1s, fell from several thousand dollars to less than a thousand on the used market, with new prices declining substantially as well. DSL carrier gear prices fell dramatically too: digital subscriber line access multiplexors (DSLAMs) that were selling new for more than $100,000 apiece in 1999 were available in surplus for about $12,000 by 2003; even new, prices fell to less than half of earlier levels.

So for the telecom equipment manufacturers, the meltdown led to a multipronged attack on their businesses. They had invested in new capacity to meet demand that only lasted for a short time. Too many new competitors entered some sectors. The total market for telecom gear then fell to low levels, even lower than before the boom. Finally, surplus equipment from the boom was available to meet much of the demand. Recovery only slowly began by 2004, in large part because the leftover equipment was getting old.

Endnotes

[1] An early term for Internet was *catenet*, apparently first used by Pouzin in a 1974 paper cited in 1978 by ARPA's Vint Cerf in IEN 48, *The Catenet Model for Internetworking.*

[2] Among the better books that discuss the Internet's early history are *Where Wizards Stay Up Late: The Origins of the Internet* by Katie Hafner and Matthew Lyon and *Inventing the Internet* by Janet Abbate.

[3] From a collection in *Mappa Mundi*, http://www.mappa.mundi.net/maps/maps_001/.

[4] Technically, X.25 defined the interface *to* the packet-switched network; the technology within a network was transparent, and indeed differed between vendors. Recommendation X.75 defined interfaces *between* networks, enabling intercarrier connectivity.

[5] Grier, D. A., George Washington University, and M. Campbell, University of Wisconsin, Madison, *A Social History of Bitnet and Listserv, 1985–1991*, IEEE Annals of the History of Computing, http://www.computer.org/annals/articles/bitnet.htm?SMSESSION=NO.

[6] The session layer was primarily concerned with maintaining the half-duplex nature of some connections, based on the way some IBM keyboards were locked when not invited to transmit; the protocol itself was quite complex. The presentation layer was concerned with negotiating the way data was encoded within the application but, as such, did not belong in a separate layer.

[7] Contrary to campaign propaganda, Gore never claimed to have invented the Internet. As senator, though, he was active in securing funding for the NSFnet, which became the Internet backbone. He also led the effort for the High Performance Computing and National Research and Education Network (NREN) Act of 1991, which provided additional Internet funding.

[8] "NSFnet "Privatization" and the Public Interest: Can Misguided Policy be Corrected?," Cook Report (1992), http://www.cookreport.com/p.index.shtml.

[9] The name was apparently a play on its ability to interconnect commercial users with research and educational (R&E) users.

[10] Cook Report (1992), http://www.cookreport.com/p.index.shtml.

[11] Exactly who was a Tier 1 provider was never entirely clear, but it was for a time typically said to be the "Big Five": UUNET, Sprint, MCI, BBN, and ANS. Some smaller early networks, such as AGIS and NetRail, also had Tier 1-level peering agreements.

[12] The author's working definition of Tier 2 network was "any network that advertises itself as Tier 1." The real Tier 1s never had to say it.

[13] Olavsrud, T., "Cable and Wireless Buys Exodus," Internetnews.com, Dec. 3, 2001, http://www.isp-planet.com/news/2001/cw_011203.html. Cable and Wireless also acquired Exodus competitor Digital Island, but itself went bankrupt in 2003, its assets being acquired by backbone ISP Savvis.

[14]   http://www.why-not.com/company/stats.htm.

[15]   Ameri, G., "The Evolution of the ISP Market as a Model for Public Wi-Fi," Telephony-Online.com, May 23, 2003, citing *Boardwatch Magazine* and eTinium.com, http://www.telephonyonline.com/ar/telecom_evolution_isp_market/.

[16]   Federal Communications Commission, *Trends in Telephone Service 1999.*

[17]   Faber, D., "The Rise and Fraud of WorldCom," Sept. 8, 2003, http://www.moneycentral.msn.com/content/CNBCTV/Articles/TVReports/P60225.asp.

[18]   Dreazen, Y., "Fallacies of the Tech Boom," *Wall Street Journal*, Sept. 26, 2002.

[19]   *Discount Long Distance Digest #59,* August 11, 1995.

[20]   *Discount Long Distance Digest #59,* August 11, 1995.

[21]   Wild, R., *Management by Compulsion: The Corporate Urge to Grow*, New York: U.S.A. Houghton Mifflin Company, 1978.

[22]   Lazarus, D., "BT-MCI Merger Alive, But for Less Cash," *Wired News*, August 22, 1997.

[23]   Cook, G., "Journey to the Center of the Internet," ISP-Planet, May 24, 2004, www.isp-planet.com/perspectives/2004/cook_internet.html.

[24]   The author was among them at the time.

[25]   Tice, D., "Credit—Not Productivity—Is the Horse We Rode In On," Prudent Bear, December 2000, http://www.prudentbear.com/bc_library_RR_onwallstreet.asp?stage=3&content_idx=8451.

[26]   Another good discussion of Teligent's fall was Dan O'Shea's "Murder by Numbers," Telephony, Oct. 8, 2001, http://www.telephonyonline.com/ar/telecom_murder_numbers/.

[27]   Edelman, R., "Tulip Bulbs and the Stock Market," Inside Personal Finance, http://www.ricedelman.com/planning/investing/tulipbulbs.asp.

[28]   The DNS was first described by Paul Mockapetris in IETF RFC882, *Domain Names—Concepts and Facilities*, November 1983.

[29]   The http://www.readme.now file from "Mosaic Netscape 0.9 beta (Windows)," dated Oct. 12, 1994, also had the footnotes, "Yes, this is Mozilla."

[30]   Duffy, T., "Market Capitalization Quickstudy," *Computerworld*, Dec. 6, 1999, http://www.computerworld.com/news/1999/story/0,11280,37807,00.html.

[31]   A few analog switches, mainly of the 1AESS variety, remained in place well into the new century.

[32]   Mann, B., "Lucent's CEO McGinn Removed," The Motley Fool, Oct. 23, 2000, http://www.fool.com/news/2000/lu001023.htm.

5

The Deuteronomy Networks

The Internet boom created a multifaceted stimulus to the telecommunications industry. On the one hand, it created a huge demand for bandwidth, leading to the exhaust of intercity fiber-optic routes that had been built during the preceding decade. It also created demand for local bandwidth to alleviate the "last mile" pressures resulting from traditional local telephone company rate structures, with their voice orientation. And equally importantly, it created a demand for investment opportunities. As a booming sector of a fast-growing economy, telecommunications was attractive to individual and institutional investors alike, in both debt and equity markets. And where capital is available, it is always easy to find someone willing to take it.

Even before the height of the boom, capital availability was improving for many telecommunications projects. During the 1980s, MCI, for instance, had been highly dependent on high-yield debt (junk bonds), as the competitive long distance sector was not yet profitable, but by the 1990s a more competitive marketplace was shaping up, and investors took notice. Still, a number of major initiatives were largely handled via private capital. The competitive access providers (CAPs), for instance, which provided fiber-optic last-mile bandwidth to business customers even before the Telecom Act opened up local competition, were financed by large corporations. Merrill Lynch helped finance Teleport, the first CAP, in order to reduce its own telecommunications costs; it was later owned by a group of cable television operators that also provided it with rights-of-way and other operational services. Metropolitan Fiber Systems, another CAP, was largely owned by Peter Kiewit Sons' Inc., a large construction firm. A number of large corporations also invested in long distance networks that were eventually rolled up into WorldCom.

The players that built the costliest new networks in the 1990s were not newcomers to the industry. They had a track record. They had generally done what venture capitalists most like to see done—built *and sold* a company. Running it at a profit was not the type of track record that counted; selling out quickly was the more favored course. Thus, a group of networks sprang up that can be called the "Deuteronomy Networks" because, like the biblical book that is largely a second telling of the law, these networks were a second telling of their founders' "story." Unlike television viewers, investors often prefer reruns.

The Short-Term Bandwidth Crunch Invited More Suppliers

As the ISP business grew and bandwidth became scarce, the market became ripe for an additional provider of long-haul service. AT&T was the largest provider of interstate voice service, but it was not a major player in the bulk bandwidth market. Its prices were too high to compete, and its network's capacity may have been too low. AT&T's network was a legacy of its ownership of the Bell System. Its rights-of-way were a hodge-podge, often shared with local Bells, and thus for it to pull additional fiber would have been particularly difficult. AT&T also suffered from being an early adopter of fiber optics: It had built much of its network in the late 1980s, before higher-bandwidth SONET equipment had come on the market. So it rationally stuck to its knitting, focusing on higher-value voice services, such as 800 service [1] numbers, which had once been an AT&T monopoly and remained a growth market after the introduction of toll-free number portability. MCI and Sprint were both heavily invested in SONET.

The original fiber-optic networks pulled by AT&T, MCI, Sprint, and Wiltel were being marketed in the usual manner, as *services*. Subscribers could get a voice-grade channel, DS1 "high-capacity" channel, or a DS3 channel. But they were not in the fiber-optic business. Of course neither were the RBOCs; the mere thought of leasing dark fiber to subscribers was enough to put the average Bellhead into fits of apoplexy. Fiber optics was a tool that these telecommunications service providers used to create their services. Sure, Sprint advertised its fiber network and even changed its stock ticker to FON (a double entendre for "fiber-optic network" as well as "phone"), but that did not mean it was in any hurry to lease the fiber.

ISPs Wanted Dark Fiber

The Internet service providers were not like most telecommunications customers. For one thing, their bandwidth requirements were simply not in the same league. For another, they were usually run by technologists who knew quite well how the telecommunications networks operated. They heard the hype about fiber optics and bought in: They wanted fiber, not just fixed bit rate services. They wanted to be able to upgrade their networks rapidly, on their own: Even though their traffic was not really doubling every hundred days, it was growing rapidly.

But it is not practical for many companies to lay their own dark fiber along the same rights-of-way. Most of the cost of fiber optics is in the installation, after all. Underground construction can cost $50,000 to $100,000 per mile. Aerial construction can cost about $10,000, if the poles are not too crowded, but poles can get crowded very quickly. It again has the characteristics of a natural monopoly. Carriers understand this. They often collaborate among themselves. Most undersea cables, for instance, are shared among many carriers. They own percentages of the cable, entitling each to a proportionate share of its capacity. Fiber-optic networks too are sometimes shared using a mechanism known as the indeasible right to use (IRU), typically for a certain number of strands along the route. A company purchases an IRU from the fiber's owner and gets total use of that fiber, typically for 20 years. The purchaser gets the dark fiber, and lights it however it sees fit.

Traditional telephone carriers did not want to sell IRUs to potential customers, but ISPs did want to buy them. And some long distance companies, which owned their own switches, were interested in procuring IRUs for their backbone transmission, again an alternative to paying for, say, DS3 wholesale circuits. This provided an opportunity to companies that could get the capital and rights-of-way to lay in a nationwide fiber-optic backbone. Without their own traditional business to jeopardize, they could sell IRUs and thus share the risk of building new networks.

Qwest Follows Sprint's Lead Along the Rails

The first of these new nationwide networks to be built was Qwest. The company was begun by Phillip Anschutz, who had, in the 1980s, purchased the Southern Pacific Railroad. The railroad itself had a history in telecom: Sprint began as the Southern Pacific Communications Company, which had built a microwave network along the railroad's rights-of-way. Many railroads owned their own microwave transmission facilities for internal use; Sprint was the first one that became a competitive long distance carrier after that opportunity arose. But Sprint was acquired by GTE, and later by United Telecom (today's Sprint Corp.), before 1988, when Anschutz bought the railroad. Anschutz only kept the railroad for eight years, but when he sold it to the Union Pacific, he kept the rights-of-way to lay fiber along the tracks.

Qwest refined the art of trenching fiber-optic cable alongside railroad tracks. It planned an initial network of approximately 13,000 route miles of 96 strands of the latest dispersion-shifted fiber, [2] and in 1997 the company hired Joseph Nacchio, a senior AT&T executive, as CEO. Qwest sold an IRU for 48 strands of its fiber to Frontier Communications for an undisclosed price, and it sold an IRU for 24 strands to GTE Corp. for a sum reportedly just over half a billion dollars. So before the first strand of fiber was actually lit in late 1997, most had been presold as IRUs, covering much of its construction cost. By 1998, as the network was coming on line, Qwest's market capitalization had reached $6 billion. By 1999, Qwest had purchased RBOC US West, and its market capitalization exceeded $65 billion. Frontier had also been acquired, by international carrier Global Crossing, which sold off the local exchange networks, and the Frontier name, to rural LEC chain Citizens Communications.

For a short time during the boom, Qwest looked like a clear winner. By purchasing US West, albeit the smallest and weakest of the RBOCs, Qwest had purchased a cash engine; even a weak ILEC has monopoly power and thus generates consistent profits. In essence it had monetized the speculative value of its stock. This proved to be very useful going forward, as increasing competition and the failure of the Internet sector led to major losses in the long distance arena. That did not, however, save Qwest when in 2001 it was found to have engaged in creative accounting practices. These practices included swapping assets with other carriers and recording them as sales (hollow swaps) and improperly accounting for IRUs. Qwest restated its sales downward by about $2 billion in 2002. Its partners in crime, Global Crossing and Enron, were also brought down by financial scandal. Qwest did emerge under new leadership and remained a major player, but its original investors were severely harmed along the way.

Kiewit Sells MFS, Creates Level 3

MFS Communications, originally Metropolitan Fiber Systems, was founded in the late 1980s and had gone public in 1993, just as the Internet was itself becoming available to the public. Its primary owner was Peter Kiewit Sons', the Omaha-based construction giant. In 1995, MFS acquired UUNET, the largest backbone Internet service provider. In 1996, WorldCom paid $14.3 billion to acquire MFS.

Having struck gold in telecom once, Kiewit promptly set out to do it again. Its subsidiary, Kiewit Diversified Group, began creating a long-haul network operation and was later renamed Level 3 Communications [3]. Its mission was to build a nationwide fiber-optic network oriented toward Internet Protocol rather than focused on the declining revenue area of traditional telephony. This not only avoided association with a sector that was already distressed, but it associated the start-up with the boom, making it, in effect, a large, self-provisioned wholesale ISP. Level 3's plan was announced in early 1998 and turned into a separate public corporation within months. Shortly after that, its IPO "successfully issued the largest junk bond offering of the 1990s" [4], raising $2 billion in April 1998. Level 3 pieced together a network using, among other things, rights-of-way obtained from Union Pacific Railroad (which in 1995 had acquired Southern Pacific) [5] and capacity obtained from Frontier, which of course owned part of Qwest's capacity. It also built local networks in 27 cities in the United States.

In 1998, Level 3 purchased XCOM Technologies, a Cambridge, Mass.-based CLEC, for $154 million. This was a huge sum for a start-up CLEC with only 10,000 lines in service, but XCOM was the pioneer of "switchless" modem operations. It had created a gateway between its Ascend TNT remote access servers and the Signaling System 7 network, enabling calls to go directly to the TNTs without going through a separate central office switch, thereby reducing the cost of providing "offload" service to ISPs. XCOM was planning to roll out similar services elsewhere; Level 3 had the financial wherewithal to do so. Investing heavily in Lucent (which had bought Ascend) remote access servers and gateways in the United States and equivalent Nortel gear in Europe, Level 3 became a large wholesale Internet access service provider. Of course this too became a crowded market and declined in profitability once the ILECs got the FCC and many states to go along with reducing the reciprocal compensation payments owed to CLECs for terminating ISP-bound calls. But "managed modem revenues" ended up, by 2004, as its most lucrative operation, representing a larger percentage of revenues than long-haul transport or IP services.

Level 3 became a survivor, but not because it was profitable; rather, it had taken in enough cash during the boom to survive for a time and continued to attract new financing, albeit at much reduced share prices. Among the later investors was Berkshire Hathaway Corp., Warren Buffet's investment vehicle, which was famously skeptical about boom investing. Level 3 was also able to take advantage of the bargains that post-meltdown companies left behind and thus improve its bottom line. But along the way, some mighty strange deals occurred. It acquired two major software distributors, Software Spectrum and Corporate Software, whose business had little in common with the rest of Level 3's operations but whose cash flow was high—a combined annual revenue of more than $2 billion. It has been suggested that the deal was done to meet financing covenants that had promised unrealistic revenue growth. If Level 3 could not grow its revenues organically, it could do so by buying profitable businesses—in other industries.

The company's status in the ISP world was elevated when, in 2002, it acquired the assets of bankrupt backbone ISP Genuity. That company was created when Bell Atlantic acquired GTE Corp. in 2000 and had to spin off its backbone assets because the company did not yet have its authority, under

Section 271 of the Telecom Act, to offer interLATA service. Genuity's IPO, which raised $1.9 billion, occurred at the trailing edge of the boom. The company had a unique management structure that gave "supermajority" power to Verizon even though the latter's nominal ownership interest of 9.5% was below the 10% cap in Section 271. It also permitted Genuity's losses to be kept off of Verizon's books, in contrast to how, as GTE Internetworking, its losses had harmed parent GTE Corp.'s bottom line. Verizon retained an "option" to reacquire Genuity within five years of the IPO, provided that it acquired Section 271 authority in most states. That option would have, if exercised, created enough shares to give Verizon a majority interest. Thus Genuity's true market capitalization could be factored in two ways, with or without the options counted in the share total; the IPO value could have been much less than the actual value had the company not failed. Even with its strike price of zero, Verizon did not choose to exercise the options. After blowing through several billion dollars in a few years, Genuity's assets, including a nationwide IP network and many high-end customers, were picked up by Level 3 for a mere pittance, $242 million. Still, Level 3 was far from profitable.

Williams Sold Wiltel, Created Another One

The Williams Companies, a large natural gas pipeline operator from Tulsa, early on recognized the value of its rights-of-way to the telecom industry. In the early postdivestiture years, it began pulling optical fiber through decommissioned pipelines. This led to the first Wiltel, a nationwide wholesale fiber-optic network that operated as a "carrier's carrier." In 1991, Wiltel introduced the first common carrier Frame Relay service, an economical alternative to private lines for private data networks. But the bulk of its business was DS3 circuits to customers such as MCI. At one point a major provider of operator services became indebted to it; Wiltel acquired the company, putting it into the retail service business. But in early 1995, before it became a household name, it was acquired by LDDS, which then renamed itself WorldCom. Williams, though, did not entirely give up its telecom ambitions. Although it entered into a three-year noncompete agreement with WorldCom, it retained its rights-of-way and kept a token strand of fiber along its routes. It also retained ownership of Vyvx, a small carrier specializing in video distribution. It kept ownership of the Wiltel trademark, which in 1997 it used for a PBX-servicing business that it acquired from Nortel.

On the day the noncompete agreement with WorldCom expired [6], Williams announced that it was back in the long-haul carrier business. The new Williams Communications had fiber optics along what were by then well-trodden routes. And because it was already 1998, Williams was a bit late to the party. Qwest was well along on construction of its network. Level 3 was starting up at roughly the same time but with an even more ambitious plan.

Williams Communications was never a top-tier player, though its network gained a fair number of customers, especially other carriers. In 1999, it sold an interest in itself to SBC, which had national ambitions; SBC would thus use Williams' network for its own branded services. Its 32,000-mile backbone was complete by the end of 2000, and Williams spun the company off to its shareholders. But in 2001, it started spinning off operations, such as its enterprise services unit. And as the company continued to lose money, CEO Howard Janzen tried to downplay the glut of fiber that had developed. "You won't find many industries that can match the growth we're seeing as new broadband applications drive demand," noted Janzen. "I believe we'll find marketplace fears about bandwidth glut are unfounded. There is strong and continued growth in the Internet and data traffic market and recent studies by industry analysts cite Internet growth rates approaching 100 percent per year" [7]. That quarter, the company lost $250 million on revenues of $283 million. It filed Chapter 11 bankruptcy in April 2002; that quarter's losses totaled $293 million. Its shares were cancelled as it emerged from bankruptcy later that year, renamed Wiltel Communications; it was then acquired by Leucadia National Corporation, a diversified investor that had acquired 44% of the company some months earlier. The total value of the stock swap was less than half a billion dollars.

Metromedia Sold Cellular and Long-Haul, Created MFN

The Metromedia name has a long and storied history. It has not so much referred to a corporation as to a single person, John Kluge, and the many enterprises he has operated. Kluge is often listed as one of the world's richest people. For many years, starting in 1946, his privately held company focused on broadcasting, owning among other things a New York City television station. He sold the station to Rupert Murdoch when the latter founded the Fox television network. Metromedia was an early investor in cellular telephone service and owned the Cellular One name, which it licensed freely to other "A-side" operators. Metromedia Cellular's East Coast network was acquired by SBC and became a major part of its Cingular Wireless network. Both of those transactions netted good profits for Kluge, who recognized the value of entering the market early. Metromedia even once had a long distance network that had been acquired by the company that became WorldCom. The post-2000 Metromedia was a different firm indeed: It evolved into the Metromedia Restaurant Group, owning such chains as Bennigan's and Ponderosa. But in the interim, Metromedia poured money into a deuteronomy network of its own.

Unlike Level 3, Wiltel, and Qwest, Metromedia Fiber Network (MFN) did not create a nationwide fiber-optic network. Instead, it created a competitive access provider, whose mission was to provide fiber-optic local loops in metropolitan areas. Unlike earlier providers that rationed out bits per second from their multixors, MFN was happy to provide dark fiber: Many of its buried pipes had 288 strands, more than enough to go around. Metromedia did not actually start the company; it was founded in 1993 as National Fiber Network by Stephen Garofalo, an electrical contractor, but Metromedia bought a controlling interest in 1997. This deal put Metromedia back into the telecom business, right when the boom was in full swing and local telephone competition was exciting investors. In 1999, Verizon (then still called Bell Atlantic) purchased a 20% interest in the company in a deal worth slightly more than $2 billion.

The problem with MFN was that the customers were not there. Trenching hundreds of miles of city streets was expensive. The ILECs were already there, of course; although they did not willingly provide dark fiber, they were required to provide "spare" dark fiber capacity to CLECs at prices based on the FCC's Total Element Long Run Incremental Cost (TELRIC) methodology, an incremental cost basis that undercut the CAPs. This was generally good for competition and for the CLECs' customers, but it was no help to the CAPs that had in many cases followed each other trenching the same major city streets. MFN filed for bankruptcy in May 2002. It emerged over a year later as AboveNet Inc., taking the name of an ISP that it had acquired earlier. As with many other bankrupt companies, it emerged with new shares and new capitalization but with a market capitalization only a fraction of its predecessors' peak value.

XO Communications Recycles Cellular Profits

Craig McCaw took his family's wireless business and grew it into one of the country's largest cellular telephone operators. In 1994, he sold McCaw Cellular to AT&T for $11.5 billion. He did not totally exit the wireless business; he soon invested more than a billion dollars in Nextel. But he also entered the wireline business with a company originally called NEXTLINK Communications. During the boom years, NEXTLINK built a competitive local network, pulling fiber in numerous cities, especially second-tier ones that competitors such as WorldCom and AT&T had not reached. It became the largest holder of LMDS [8] fixed-microwave licenses from the FCC, enabling it to build high-capacity point-to-point wireless links. In 2000, it merged with ISP Concentric Communications and renamed itself XO Communications.

In 2001, the company arranged an $800 million infusion from two major investors, Forstmann-Little and Teléfonos de Mexico, but this was just the beginning of its financial "restructuring" that washed out most of its original investors' capital. Those two investors backed out. XO filed Chapter 11 in June 2002 and its primary debt holder, investor Carl Icahn, ended up in control. Icahn had avoided the boom and instead concentrated on buying companies in distress; XO was the kind of bargain he liked. The company resumed public trading on the NASDAQ and, in 2004, acquired bankrupt CLEC Allegiance Telecom. Along the way it missed out on attempted purchases of Global Crossing and Britain's Cable and Wireless. Without quite achieving profitability, XO had morphed from a boom tycoon's dream into a bottom-feeding tycoon's vehicle for acquiring cheap assets.

Undersea, Undersea, Under Beautiful Sea

The undersea cable business has long been a rather special part of the world telecom industry. Requiring massive investments, undersea cables have traditionally been owned by consortiums of major carriers. For example, in 1992, TAT-9, a fiber-optic cable carrying 560 Mbps, was installed between the United States and Europe. It cost $450 million and had 39 owners, including AT&T, British Telecom, Telefonica (Spain), Teleglobe (Canada), and France Telecom [9]. This was the normal business model; carriers received capacity in proportion to their investment.

But during the boom, undersea cables became another area for investors to squander their surplus cash. In 1997, Atlantic Crossing was proposed by newcomer Global Telesystems Ltd. as a 40-Gbps fiber link between the United States, the United Kingdom, and Germany. By the end of the year, the company had renamed itself Global Crossing Ltd. and added a cable to Japan (Pacific Crossing) and one to the Caribbean (Mid-Atlantic Crossing) to its plans. By 1998, the company, nominally based in the tax haven of Bermuda but actually run from New Jersey, added plans for Pan European Crossing, a terrestrial network among many of Europe's largest cities. In 1999, it announced plans for additional fiber in Asia. It also tried to purchase RBOC US West but lost out to fellow boomer Qwest. As a consolation prize, it acquired Frontier Communications, which had a second-tier domestic long distance network (originally called Allnet) as well as a mid-sized independent ILEC. After Global Crossing sold the ILEC business to Citizens Communications, it kept the fiber, which of course was largely in the form of Qwest IRUs. All of this ambitious growth came at a cost, of course. Had its sales projections been correct, it might have worked. But then whose were? Chapter 11 followed. The company was acquired in 2003 by Singapore Technologies Telemedia for $250 million.

FLAG Telecom (named for "fiber link around the globe") planned to spend $1.5 billion to lay 27,000 miles of undersea cable. More of a carrier's carrier than Global Crossing, it was more speculative than previous consortium cables. Its largest owner (38%) was Bell Atlantic, which in 1996, when FLAG was announced, was far smaller than the company it grew into as Verizon. Its planned route, which was to have 8 Gbps of capacity, went mostly undersea from England to Japan by way of Egypt, Dubai, Thailand, and China [10]. Some of FLAG's network did get built, but it went bankrupt in 2002 under $2.5 billion of debt.

Other ambitious undersea networks were planned during the boom, some getting further than others. The biggest plan was called Project Oxygen. Little-known CTR Group announced it in 1997. The network was to have 320,000 kilometers of undersea cable touching 171 countries. It never got the $10 billion that it needed, though. Dreams like this required deep pockets, since undersea cable can be extremely costly. One description goes like this: The cost of a trans-Atlantic cable is roughly the cost of that length of $10 bills, rolled up end to end. To cross the Pacific, substitute hundreds.

As a result of the glut of undersea bandwidth, costs of transoceanic calls plummeted. Bulk bandwidth between the United States and Europe became inexpensive, and even Asian routes became far more affordable. Who would have expected that some flat-rate North American calling plans would throw Hong Kong in to the bargain? By 2004, a New York to London STM-4 [11] circuit (622 Mbps) was being offered at $7,500 per month. That is more bandwidth than the entire TAT-9 of a dozen years earlier! The original investors in the cables, however experienced they might have been in other areas, were not the ones making the money.

How Much Bandwidth Was Available?

The *potential* capacity of fiber-optic networks has increased dramatically since the original transcontinental fiber networks were laid in the 1980s. In 1985, it was common to install 135-Mbps fiber, capable of carrying three DS3 signals; this was a huge improvement over earlier systems, and it was adequate for the day's applications. The highest-capacity trunk routes required multiple strands, but extra strands were always pretty cheap, compared with laying the pipe in the ground in the first place. High-end systems of the day operated at up to 565 Mbps, trumping the capacity of any earlier coaxial cable or microwave radio system.

SONET was standardized in the 1980s and first hit the market around 1990. It was very important to the RBOCs: Earlier systems were proprietary, so a given route was locked in to a single vendor's gear. SONET was a standard that allowed multiple-vendor interoperability and thus allowed for more competitive bidding. The first round of SONET gear installed in the early 1990s ran at OC-48 (2,488 Mbps). That was technologically exotic and difficult to implement, given that the microprocessors of the day only ran at about 2% of that speed.

But as semiconductor and optical technology improved, faster speeds became possible. The next major speed bump in SONET was to OC-192, almost 10 Gbps. And OC-768, near 40 Gbps, has been produced. These speeds do not run reliably, at least not very far, on some of the oldest single-mode fibers, but the boom years were accompanied by improvements in fiber itself, so these speeds are achievable.

But that is not the only way to increase fiber capacity. Because the semiconductor lasers used for fiber optics transmit on *precisely* a specified wavelength ("lambda") [12], it is possible to have multiple wavelengths of light share a single strand of fiber. In the earliest systems, it was possible to have two, four, or maybe eight lambdas. This WDM was the trick that everybody had up their sleeve. But as the boom moved along, WDM technology improved. Optical engineers were able to position the lambdas very close together—100 GHz apart, in some systems—such that a single fiber could carry several dozen lambdas. This came to be known as dense wavelength division multiplexing (DWDM). Investors rushed in, buoyed by forecasts of massive Internet growth. But of course the demand did not happen. Few carriers actually *needed* DWDM. Most of the fibers in the ground still were not lit with even a single lambda.

The upshot was that by the early 2000s, the industry began to emphasize "coarse" WDM (CWDM), with only a few lambdas on the fiber. CWDM is cheaper to implement and adequate for most users. In the unlikely event of a fiber shortage, DWDM technology remains on the shelf.

A Falling Price Lowers All Carriers' Ships

Needless to say, the price of bandwidth has been falling rather steadily. This is most pronounced at the wholesale level. Retail long distance prices have a dynamic of their own. Low-volume callers are not terribly profitable. They are most likely to get service from their local exchange carrier, now that all of the RBOCs have the necessary Section 271 authority [13]. But wholesale prices have been impacted dramatically.

By 2004, Verizon became the second-largest long distance carrier, in terms of number of subscribers. But its average per-subscriber revenue was low. AT&T, MCI, and Sprint concentrated on larger business customers and 800 service. AT&T's 2003 revenue was almost 9% below its 2002 revenue, whereas Sprint's wireline revenue—a mix of local and long distance—was down 1.6%. MCI's was down 10%. This was not just attrition of customers to ILECs; it was the result of a general fall in prices. By 2004, leased OC-3 circuits (155 Mbps) were available between New York and Los Angeles for $5,000 per month.

Twenty years earlier, a single DS1 circuit on that route, 1/84th of that capacity, was about $60,000 per month. For most of the intervening years, increased supply was met by expanding demand, leading to increased overall revenues. But since the meltdown, a glut has led to falling revenues, with no end in sight.

Endnotes

[1] Toll-free inbound calling was originally called "Inward WATS," then renamed "800 service." This name is a bit anachronistic, of course, in light of the additional service access codes in the 8nn range that subsequently have been assigned to it.

[2] Steinberg, S., "Crucial Tech: Telecom Goes Qwest,", *Wired*, March 31, 1998, http://www.wired.com/news/technology/0,1282,11371,00.html.

[3] Level 3 did retain some of Kiewit Diversified Group's other assets, ranging as far afield as KCP Inc., a coal mining firm.

[4] Level 3 press release, "Level 3 Sells Junk Bonds," April 27, 1998.

[5] Level 3 press release, "Level 3 Communications and Union Pacific Railroad Sign Fiber Optic Right of Way Agreement," April 2, 1998.

[6] Williams press release, "Williams Returns to its Roots With Launch of Wholesale Network Services for Nationwide Market," January 5, 1998.

[7] Williams press release, "Williams Communications CEO Refutes Bandwidth Glut," June 12, 2001.

[8] LMDS operates in the 27–32 GHz band, where the typical range is about two miles but link capacity is competitive with coaxial cable.

[9] AT&T press release, March 2, 1992.

[10] Stephenson, N., "Mother Earth Mother Board," *Wired*, October 1996.

[11] The North American SONET hierarchy measures bandwidth in "STS" capacity units of 51.84 Mbps, which are equal to one optical unit (OC). The European SDH hierarchy, *almost* the same in many technical respects, measures bandwidth in "STM" units of 155.52 Mbps, each equal to three OCs. Thus, STM-4 is the same bit rate as STS-12, also called OC-12.

[12] Wavelength, of course, is the inverse of frequency; the optical spectrum is simply a higher part of the electromagnetic spectrum than radio waves. We perceive optical wavelengths as different colors, but fiber optics perform best in the infrared range.

[13] That is, approval by the state and FCC according to Section 271 of the Telecom Act, which releases them from the MFJ prohibition against providing interLATA services.

6

Losing by Winning: Wireless License Auctions

Although the move from monopoly to competition has arguably been the major force reshaping the telecommunications industry since the divestiture era, the most *visible* change may well have been the growth of wireless services. Mobile telephones were a rarity in the early 1980s; today, in many countries they outnumber wireline phones, and even in the United States they represent a large share of the voice market. This growth has primarily been the result of technological advances, but regulators worldwide have accommodated it by spectrum allocations and regulated network interconnection.

But the rapid growth of the wireless communications industry, especially during the boom years of the late 1990s, led to a string of events in which both the regulators and the regulated lost sight of the business imperatives and managed to find financial hardship amidst unprecedented success. Major wireless carriers around the world, mostly affiliated with major wireline carriers, bid amazing sums at auctions for new wireless licenses. This was good for taxpayers but left the auction winners saddled with unsustainable debt. How did it come to this?

Wireless [1] telecommunications in the 1970s was a niche market. Two-way mobile voice communications was primarily limited to private systems, generally known as *land mobile* radio. Motorola had the lion's share of the equipment market, selling to public safety agencies such as police and fire, as well as utilities, taxicab fleets, and other industries that needed instant communications with a mobile workforce. Companies that needed systems generally had to buy everything themselves. They might put up a tower for the base station or rent space on a tower or other high point. The base station might be a half-duplex [2] dispatch radio or a full-duplex *repeater* that enabled all units within its service range to talk among each other. Demand for these systems grew continuously in the years after World War II, and the FCC routinely added new frequency bands for them, but frequencies were still usually tight in major markets.

Then there were the radio common carriers (RCCs). These companies generally operated one-way paging systems for the public. They rented out "beepers," which would respond, when signaled by a high-powered radio transmitter, when a person dialed the beeper's assigned telephone number. The RCCs' customer base included doctors, volunteer firefighters, and others who needed to be reached at all hours. Beepers were generally one-way devices, sometimes capable of displaying a return phone number and occasionally able to broadcast a brief voice message. But they were not mobile telephones; the recipient of the beep typically had to scurry to the nearest pay station to respond to the message.

Car phones did exist, but they were very rare. The original car phone system in the United States was known as MTS, for mobile telephone system; it used full-duplex VHF FM radios and needed operators to place calls. Bandwidth capacity was very limited; only a few frequency pairs were available in any given city, so only a few hundred MTS phones could be issued by the local telephone company. This was followed up by IMTS (improved MTS), which allowed direct dialing of calls; IMTS phones actually used rotary pulse signaling to dial digits. But that did not solve the frequency availability problem; IMTS phones were generally the province of tycoons and others who could afford to wait years for one to become available.

In the early 1970s, two companies raced to be the one to make mobile telephones more available, though neither imagined how successful they would become. AT&T, then owner of the monopoly Bell System local telephone companies, sought to make car phones more available than IMTS could be. Motorola, then the leader in hand-held radio manufacturing, set out to make a hand-held mobile phone. In 1973, Motorola's Martin Cooper made the first call on a 30-ounce hand-held prototype of the first cellular phone, and although the 10-inch-high handset was a bit heavy, further development led to a continual reduction in size and weight. Cooper is widely credited with inventing the cell phone. But it was not just the radio components that mattered. The mobile phones themselves operated under the command of the network, following a complex protocol, shifting frequencies on command. Was it entirely a coincidence that Motorola's semiconductor division introduced its first microprocessors just after Cooper's first cell phone was built?

The system is called *cellular* because it divides the network's service area into small geographic areas, each served by a base station, and as traffic grows, capacity can be increased by *splitting* these *cells* into even smaller areas, adding more base stations and reducing the area covered by each. Thus instead of using a single radio frequency-pair only once within a given market area, a frequency pair can be reused in different cells. In the prototypical layout, cells are arranged in a hexagonal grid pattern, each cell surrounded by six others using different frequencies, allowing a nominal 7:1 frequency reuse pattern. Thus a network operator only needed seven sets of frequency pairs and could cell-split its way to ever-increasing capacity.

Well, that was the idea, but in reality it was not quite that easy. The first cellular telephones were based on FM analog radio transmission, and radio waves do not stop on command. Cell splitting works only so far, and cell sites are costly to construct. In an urban area, cells could shrink to a matter of blocks, and it might not be possible to find a new cell site in the desired location. So analog first-generation (1G) cellular telephone networks, which began to roll out in the 1980s, became capacity limited.

The original analog system in the United States was called Advanced Mobile Telephone Service (AMPS). European countries adapted the idea to their own available frequencies (radio frequency allocations above 30 MHz, the top of the shortwave spectrum, differ widely between world regions), resulting in several incompatible variations. So while roaming between the AMPS systems in, say, NYNEX Mobile and Franklin County Cellular in Massachusetts could result in lost incoming calls and high "gotcha" rates for outgoing calls, roaming between Germany and Denmark was not even possible using the same equipment, not that it would originally have been permitted.

Original License Lotteries Led to Farcical Resale

Before the first commercial cellular networks could be rolled out, someone had to determine who got to operate them. AT&T had originally hoped to build a nationwide system, but while the FCC was considering its application, the company broke up eight ways. In its 1984 divestiture, the existing mobile operations were given to the RBOCs, not AT&T. And the RBOCs, not AT&T, became the applicants for cellular system licenses.

The FCC recognized that competition would be beneficial for consumers, but it did not want to divide the available frequencies—originally 666 channel pairs in the 800-MHz range that had previously been television channels 71–83—too many ways. So it decided to set up a duopoly system, with two cellular licensees in each local area. One would be the so-called *wireline*, or "B-side," licensee; the other would be the *non-wireline* or "A-side" licensee. Cellular systems were still very speculative; they would require a large investment to build out, and their profit potential was by no means proven. After all, the industry at that point was largely focused on getting beyond the capacity limitations of IMTS, not in creating a mass-market product. So the initial licenses were given out in major metropolitan areas. At first the FCC tried to hold competitive hearings ("beauty contests") to determine which company would get the A-side license, but this proved to be unwieldy after 30 or so were handled this way. The remaining A-side licenses were then raffled off among existing RCCs. The B-side licensee was the monopoly local exchange carrier. If there was more than one local exchange carrier doing business within the license's footprint, a lottery would be held to select the winner.

This led to some anomalies. The Boston license area had two local exchange carriers within its footprint. NYNEX Corp. had *almost* all of the phones in Eastern Massachusetts and Rockingham County, New Hampshire. But tiny Naushon Island, off the southern coast of the state a few miles from Cape Cod, had its own little LEC, the Elizabeth Island Telephone Co. Like the privately held island itself, it was owned by the Forbes family. Its few dozen telephones were actually served by the NYNEX switch in Falmouth, but its franchise gave it equal standing in the cellular lottery. NYNEX saw fit to make a generous offer to absorb the tiny company! This did guarantee that NYNEX won Boston. But the adjacent license for Hillsborough County, New Hampshire (the state's most populous, bordering Massachusetts and including the cities of Manchester and Nashua) was won in the lottery by Continental Telephone, which at that time served a few rural towns in the county. For several years, until NYNEX purchased the franchise, the Hillsborough County line marked the beginning of *roaming* for subscribers from both sides, as the two systems were initially quite separate.

And that was another problem with the original licenses. The metropolitan area licenses were allocated based on standard metropolitan statistical area (SMSA; the term was later shortened to MSA) boundaries, creating 305 little markets. But then rural areas were allocated even more granularly, with hundreds of rural service areas (RSAs) allocated their own licenses. Sometimes an RSA was just one county; often it was a sparsely populated stretch between SMSAs. This created a patchwork quilt of licensees, each required to make its radio signals effectively stop at county lines. Needless to say this was not terribly efficient.

The Top Cellular Networks Grew to Profitability

As cellular telephony began to catch on, the licenses became more valuable. Thus the companies that owned the licenses—typically a web of subsidiaries of a larger corporation—became more valuable. These companies themselves came into play. Metromedia, for instance, had owned a major RCC that won the

A-side licenses in several cities, including Washington and Boston. It established a set of subsidiaries that shared the Cellular One trademark among themselves and with other A-side providers. In 1987, Metromedia sold its wireless holdings to Southwestern Bell (later called SBC Corp.), which held the B-side licenses in other parts of the country and was interested in expanding its footprint. Small cellular operators nationwide were cashing in, selling their systems to larger operators that wanted to expand their footprints. European countries had, as a general rule, awarded countrywide licenses and were working to harmonize their systems into a seamless mesh, whereas the U.S. model of small local systems, perhaps efficient for taxicabs and plumbers, was being revamped by the market as systems consolidated. Only it was the original cellular licensees, not the government, that were realizing the value of the licenses.

By the 1990s it was clear that cellular mobile telephony was going to be very popular. Even though systems were cash-flow negative for many years, the potential future profits became obvious. Mobile telephone systems had heavy upfront expenses. Base stations had to be built. Towers were constructed all over—this was good revenue for tower-rental companies like American Tower, which grew rapidly. Central office switching systems, known as mobile telephone serving offices (MTSOs), had to be bought—this was a good revenue source for Lucent, Nortel, Ericsson, and Siemens, among others. And the cellular phone companies generally subsidized the purchase of handsets, typically to the tune of $150 to $200 per unit, as each new customer signed up or renewed a contract. But revenues continued to rise, even as prices fell in a competitive market. The largest providers became profitable, though smaller ones, with fewer customers per cell site, never did break even.

Networks Go Digital

By the early 1990s, digital cellular systems were beginning to replace analog ones. In Europe, a multinational committee formed in 1987, originally known by its French name *Group Speciale Mobile*, created a standard that was adopted in most countries around the world. The GSM standard was later renamed "Global System for Mobile." It was a digital system, based on time division multiple access (TDMA) technology, in which individual users transmitted digitized, and compressed audio in time-synchronized time slots on a single channel. GSM was a rollicking success: It united European cellular networks under a single technology, improving production volumes of both handset and base-station equipment, enabling seamless roaming *almost* worldwide, improving security (digital signals cannot be easily listened in on using a scanner, though GSM is hardly cryptographically secure), and somewhat improving bandwidth efficiency compared with analog.

North American operators did not jump on the GSM bandwagon. In part this was because GSM was not designed for easy compatibility with AMPS. So second-generation (2G) cellular in the United States and Canada moved in different directions. Two standards initially competed for the marketplace. One system, usually known simply as TDMA though more formally as "D-AMPS" or IS-54, and later IS-136, used a different variation on TDMA technology to squeeze three calls into each original channel. This was adopted by AT&T Wireless (originally McCaw Cellular, before AT&T purchased it) and SBC's Cellular One, among others. The competing system was based on an entirely different technology, code division multiple access (CDMA), developed by Qualcomm Corp. Based on spread spectrum technology, which spreads many users' signals across a wide frequency band rather than assigning different frequencies to each user, CDMA promised greater frequency efficiency, better voice quality, and longer battery life than TDMA. But it was a bit later to market and costlier to deploy. Bell Atlantic NYNEX Mobile and GTE Wireless (both later merged into Verizon Wireless) and Sprint Spectrum were major GSM adoptees. Both of these digital systems improved network bandwidth efficiency enough to allow usage prices to go down while improving profit margins at the same time.

2G also permitted a little bit of data transmission, typically 9,600 bps. This led to a fairly widespread rollout of hand-held microbrowsers, designed for the four-to-five-line text window of the typical cell phone, using a protocol called Wireless Access Protocol (WAP). WAP was rolled out worldwide in the late 1990s. It was given marketing names such as "wireless Web," but it no more resembled the real World Wide Web than did a frozen "slyder" hamburger from the convenience store resemble a sirloin steak. WAP suffered from its small size, its limited "walled garden" content (unlike the wide-open Internet), and its by-the-minute price. Low speed was the least of its problems.

Auctions as a Fair Way to Allocate Scarce Spectrum

By the early 1990s, the U.S. Congress, like many other governments, realized that wireless telephony licenses were extremely valuable, too valuable to be given away. So Congress passed a law requiring the FCC to hold auctions for most future licenses. During the late 1980s and early 1990s, a cottage industry had sprung up in Washington helping speculators file applications for the remaining wireless licenses, which were being awarded by lottery. Lotteries were also used for other emerging radio services, such as multipoint distribution service, a fixed system originally billed as "wireless cable." Of course the idea was that the lottery winner would immediately auction off the license to the highest bidder anyway. So the FCC merely cut out the middleman.

The PCS Auction Was a Success

A seminal bandwidth auction took place between December 1994 and March 1995. The FCC auctioned off the first two "wideband personal communications service" (PCS) licenses out of the six planned for each market nationwide [3]. PCS operated at a higher frequency than cellular (1,850 –1,990 MHz, compared with 825–894 MHz for cellular), but it was still adequate for many of the same applications. PCS as originally conceived could operate in different modes—a proposed "low-tier" PCS was a sort of souped-up cordless phone, not suited for mobile use. But "high-tier" PCS was a functional substitute for cellular (a name that, in the United States, formally applied only to the 800-MHz systems) and thus appeared to have the highest profit potential. Given the imperatives of license auctions, an alternative such as low tier with its lower profit potential (even if the capital expenditure were somewhat lower, which remained to be seen) would not be economically practical. That first PCS auction, for the so-called A and B blocks of 30 MHz apiece, with a fairly large geographic license area [4], raised close to $7 billion dollars—far more than some forecasters had expected. License valuations are typically stated in terms of "$/pop," the number of dollars per person living in the license's footprint. Licenses in some top-tier markets, such as Atlanta and Chicago, went for about $25 to $30/pop. Prices were much lower in more rural markets, like Spokane, Wash., and Omaha, Neb., where they were closer to $3/pop—perhaps bidders assumed that the older cellular bandwidth would not be exhausted so soon. Major bidders included AT&T, Sprint Spectrum (jointly owned with three major cable companies), and PrimeCo, a joint venture of several RBOCs.

The PCS C-block auction, held a year later, yielded even more surprising results. The FCC had reserved two blocks, the 30-MHz-wide C block and 10-MHz-wide F block, for small businesses and other "designated entities." The idea was to reserve some spectrum for newcomers. But it was a farce—what small business could afford to participate in this type of auction? The answer turned out to be that a newly created company called NextWave bid *very* heavily. Backed by wealthy investors, including Qualcomm and Global Crossing, but technically a small start-up, NextWave bid an amazing $4.7 *billion* for licenses in 56 markets, including New York, Washington, and Los Angeles. Several of NextWave's bids went over $50/pop, often paying about three times what winning bidders paid for the same amount of spectrum the year before. And that earlier auction had been open to all bidders (see Figure 6.1).

But NextWave had another trick up its sleeve. The company was awarded the licenses after making only a down payment, about 10% of the total. (The FCC had allowed licenses to be paid off in installments, again a concession to

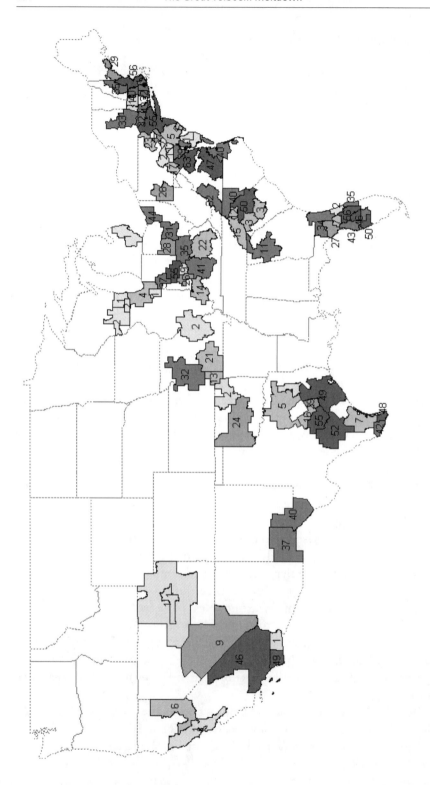

Figure 6.1 NextWave's high bids in 1996 auction of PCS C-block auction, in $/pop.

make it easier for smaller businesses to participate.) NextWave did not build anything, and it did not make any further payments. In 1998, it filed for bankruptcy, claiming the licenses as a bankruptcy asset. The FCC tried to cancel the licenses, for the simple reason that their terms (payment) had not been met. In early 2001, with boom prices still dominating the market, the FCC reauctioned the disputed licenses, raising $16 billion for them. But NextWave did not surrender and continued a court battle to hold on to the licenses. A deal was struck in late 2001 in which NextWave would get $6 billion out of these auction revenues, but Congress did not go along. In January 2003, the Supreme Court upheld NextWave's position that bankruptcy law trumped telecommunications law. But by then, the boom was over, and the market value of the licenses was going down. In 2004, NextWave and the FCC reached a settlement in which the company returned some of its licenses, kept others, and agreed to share with the government the profits of license sales or leases within a limited time. It then sold some of its licenses in a private auction that netted less than half of the voided 2001 auction's winning bids. But eight years had passed in which a major chunk of spectrum had lay fallow.

"3G" Combined the Allure of Both Internet and Wireless

With "Internet" on everyone's mind during the boom years and "dotcom" actually being viewed positively by some marketeers, the digitization of wireless telephony via 2G technology was seen by some insiders and investors as quaint, voice-centric, and thus uninteresting. Customers seemed fairly happy, at least with voice service, but equipment providers were looking for the next big upgrade to sell to their customers, the carriers, and the carriers were looking for ways to cash in on the Internet boom, which some observers still saw as being permanent.

The obvious answer was a new generation of cellular telephone technology. So-called third-generation(3G) networks would have data bandwidth competitive with wireline networks and would use Internet Protocol. A 3G system would support speeds in the one-megabit range for fixed stations, considerably slower for mobiles. But while it was easy to get an agreement on the idea of 3G, the details proved more contentious. The International Telecommunications Union created an umbrella program called International Mobile Telecommunications-2000 (IMT-2000). The GSM camp united behind the 3d Generation Partnership Project (3GPP). Qualcomm, the San Diego-based company behind 2G CDMA technology, proposed an upward-compatible upgrade called cdma2000 and was backed by a consortium called "3GPP2." Both sides agreed that the new standard should be based on some form of code division multiple access (and thus owe Qualcomm some patent royalties). But the

specific standards chosen by the GSM advocates, called wideband CDMA
(W-CDMA) or universal mobile telephony system (UMTS), was seen as unac-
ceptable by many Americans [5]. Attempts to reconcile the two resulted in a
typical compromise: IMT-2000 treats both cdma2000 and W-CDMA as 3G
standards.

Well before the standards for 3G radio transmissions were written,
equipment vendors had begun to speculate about the types of devices that
3G would support. Like the integrated voice and data workstation developers of
the early 1980s that designed (though rarely found customers for) "futuristic"
telephones-cum-computer terminals, 3G fantasists sketched a number of potential devices, typically combining the personal data assistant (PDA) with a
mobile telephone. This tended to meet with a reality problem: A good PDA
is generally too large to be a good mobile telephone; a mobile telephone's
screen and keyboard area is too small to be a good PDA, let alone computer.
A more straightforward approach would be to simply allow the user's laptop
computer or PDA to connect to the mobile telephone, which would function as a wireless modem. Or perhaps a dedicated wireless modem could be
used. Much less sexy, perhaps, but more practical. Eventually such products did
come to market, after the first 3G networks were built with little public
response. The cognitive dissonance between futuristic devices and practical
accessories was just one of the unresolved issues impeding third-generation wireless networks.

European PTTs Had Recently Been Privatized

The European mobile operators were for the most part private corporations.
Indeed by 1998, when 3G was taking off, most of Europe's wireline telephone
networks had already been privatized. Until a few years earlier, the bulk of the
telephone networks had been operated as monopolies by the PTTs. Often, the
government turned it into a publicly held corporation and sold shares, keeping
some for itself. And competition was just starting to creep in to most of Europe;
although the United Kingdom had had some competition since the early 1980s,
it was not until the boom years that the bulk of the European Union saw competitive networks develop. So management of these corporations was somewhat
new to the idea of a free market. They had never really worried too much about
money before.

Mobile telephony was different from the wireline business. It was almost
never a monopoly. The mobile affiliate of former PTT British Telecom (subsequently spun off as O2), for instance, faced strong competition from carriers
that included Orange (owned by France Telecom), T-Mobile (Deutsche Tele-
kom), and Vodafone. When the 3G license auction came along, nobody wanted
to be the one left standing after all the chairs were full.

Bubble-Era Timing Led to Spectacular Bids

The 3G auctions began in 2000. Spain and Finland had awarded licenses based on a "beauty contest," not an auction, but the United Kingdom put five licenses up for auction. With all the existing networks wanting to preserve their position and new bidders able to use this as a vehicle to enter the market, bidding was fierce. Winning bidders offered more than $100/pop for the license alone—the five British licenses together netted about $35 billion! Germany's auction, which followed soon afterwards, was similar, bringing in $47 billion. Figure 6.2 shows these bids and how they compair to others. This was good for the taxpayers, but the "winning" bidders soon realized that they had a huge monkey on their backs. Could they afford these licenses and the new networks that they were intended to support?

Other countries' subsequent auctions did not net such incredible results, as the industry's hangover began to become obvious. The Netherlands held an auction that got relatively little response—there were enough licenses to go around for the incumbents, so only about $2 billion was raised. Italy's auction raised only about a third as much as the United Kingdom's. The regulators' reaction? To investigate the bidders to see if there was collusion [6]. How could sanity break out when it was *their* turn to receive an insane windfall?

"2.5 G" Technologies Suffice for Most Users

The move to 3G mobile networking was not only stifled by absurd license auction bids but by a more fundamental business reality: There was little demand for its services that could not be met more cheaply via other means. For the

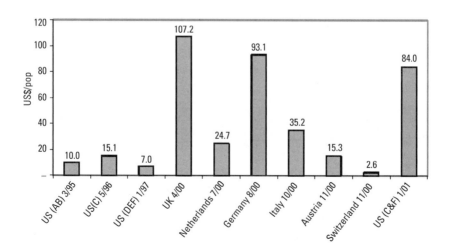

Figure 6.2   Mobile license bids in several countries during the boom (copyright Elsevier Sciences).

GSM network operators, 3G meant a whole new network: New frequencies with new base station radio equipment backed up with a new supporting infrastructure, supporting new terminals. North American TDMA operators recognized that their technology was at a dead end, and they generally looked to move ahead with the GSM world. CDMA operators had a more graceful evolutionary path but still did not need to jump all the way to full-scale cdma2000 "3G" speeds.

But then such solutions were already becoming available, and they did not require 3G. Instead, a number of less-costly technologies rolled out. These offered improved data performance, compared with plain old 2G, but without new frequencies or completely new base stations. They were nicknamed "2.5 G", the halfway step to 3G. And network operators worldwide jumped on them.

In North America, for instance, some of the original cellular operators had developed a service called cellular digital packet data (CDPD) that even worked in conjunction with first-generation analog networks. CDPD carried data, in the form of IP packets, on vacant channels in the 800-MHz cellular band. Its 19.2- Kbps burst speed was not spectacular, but it was sufficient for many applications such as e-mail and text message delivery. Although it was far from universally deployed, CDPD was installed in most major American cities. CDPD was not even 2.5 G technology, but it pioneered the cellular data market.

The GSM world has its own migration path. Beginning in 2000 or so, GSM networks have begun to roll out general packet radio service (GPRS). This adapts GSM's digital transmission scheme to allow peak data rates in the 50–100 Kbps range, using connectionless packets and thus "full-time" connectivity versus the circuit-switched model of telephone calls. Its key technical feature is the ability to access all eight time slots in a GSM TDM channel, thus theoretically allowing it eight times the burst rate of a regular call. GPRS is followed by EDGE (Enhanced Data Rates for GSM Evolution, although "GSM" was originally "Global"), which triples the data rate of GPRS by using a more advanced modulation scheme. Another approach, HSCSD, for high-speed circuit-switched data, simply takes up to four GSM telephone call time slots and aggregates them into a single data channel of up to 57.6 Kbps, provided the bandwidth is available. This lacks the full-time connectivity of pure packet approaches but allows for more predictable bulk data transfer. It has not, however, gotten as much marketplace traction as GPRS.

The CDMA world's version of 2.5 G technology is generally known as 1xRTT. This operates within the same 1.25-MHz channels as ordinary 2G CDMA voice calls, allowing packet data bursts in the 150-Kbps range. (Qualcomm positions this as "3G" technology, but that is a bit of an exaggeration; it is, however, part of Qualcomm's cdma2000 product portfolio.) CDMA networks have largely been upgraded to "1X" technology—this typically required a single board change in the base stations, not a "forklift upgrade." Full-scale 3G operation combines three channels and is called 3xRTT. Another variation, 1xEV-DO (data only), provides higher data rates within a given channel, at the expense of sharing it with voice. By 2004, 1xRTT had been widely deployed in the United States, Canada, and South Korea, the largest CDMA markets.

Many Large Incumbents Were Left With Huge Debt

The network operators that paid top dollar in the bubble-era license auctions could hardly be said to have "won" the auctions. Instead, they were saddled with such debt that they had trouble actually building out the 3G networks; indeed, the parent companies, some of them established former PTTs, suddenly found themselves mired in far more debt than they could comfortably handle. For example, France Telecom's total debt in 2004 was more than $60 billion—much of this for its forays into mobile telephony, such as its Orange network, the second largest mobile operator in Europe. And Deutsche Telekom's $65 billion debt, while counterbalanced by a somewhat stronger equity position than its French neighbor, was largely incurred by its T-Mobile operations.

Telephone companies were traditional "widows and orphans" stocks. Although such sinecures can no longer be expected in an increasingly competitive world, the wireless telephone business has been a tough one for many companies to learn. The technology came easily. The business part is still evolving. These companies have traditionally moved slowly. They were a bit slow to get on the bandwagon when the boom hit, but they were even slower to jump off when the meltdown began.

Endnotes

[1]   Even the word "wireless" carries baggage: In the early twentieth century, radio was called "wireless," as in "wireless telegraph," but that term went out of fashion by mid-century, and by the 1970s it was as quaint as a 5-cent bottle of Moxie. It came back into widespread use to refer to radio-based telecommunications services such as cellular telephony.

[2]   *Half-duplex* means that the station both transmits and receives, but only in one direction at a time, usually on a single frequency. *Full-duplex* means that it does both at the same time, requiring a pair of frequencies. A typical *repeater* installation was full duplex at the hub but with half-duplex mobile units, so a signal at the repeater's input would be retransmitted in real time.

[3]   Narrowband PCS refers to two-way paging service, capable of transmitting short text messages. Although these licenses were auctioned off and systems built, they had a much smaller impact on the marketplace.

[4]   Specifically, the PCS licenses were allocated by major trading area (MTA), of which there were 51, for the A and B blocks, and basic trading area (BTA), of which there were 493, for the C, D, E, and F blocks. These were based on a Rand McNally atlas and used

without permission, a detail that was brought to the FCC's attention somewhat later. Future auctions used U.S. government-defined geographies.

[5]   Besides its inability to interoperate with 2G CDMA telephones, the bandwidth footprint of UMTS, as originally proposed, was seen as being just a bit too wide to fit into the 5 MHz-wide channels of the American PCS bands. Europeans expected 3G to have its own bands, but the American PCS allocations used up the spectrum originally planned for 3G.

[6]   See, for instance, "Telecom Yearbook 2000: Going Going Gone!" Observatoire des Stratégies de Technologie de l' Information et de la Communication, http://www.int-evry.fr/lfh/ressources/Ostic/3G_Auction.htm.

7

Competitive Access Providers, the Costly Way to Local Competition

Although local telephone competition is often said to have begun in the United States after the passage of the Telecom Act of 1996, its roots go back more than a decade earlier. It was the AT&T divestiture, after all, that locked in the two-tier industry structure, with an ironclad wall—the LATA boundary—between local and long distance service providers. Divestiture's restrictions applied to the RBOCs themselves but not to other providers. Long distance companies could create as many Points of Presence as they wished, and they were not prohibited from directly connecting their facilities to their customers. It just was not practical very often.

RBOC Prices to Large Customers Were Out of Line

The FCC's *MTS and WATS Market Structure* decisions, which laid out the rules for the postdivestiture industry, created the access charges that interexchange carriers paid to the local exchange carriers. The newly minted RBOCs and the independent operating companies alike looked to these charges as a way to replace the heavy subsidies that had come from the historical separations and settlements process. Politically, they saw the advantage in minimizing increases in their basic monthly residential rates. So they instead looked to other sources of revenue, and the IXCs were an obvious target. The IXCs, of course, merely passed these charges along to their subscribers—in a competitive market, they were not likely to want to swallow major losses from some customers in the hope of making it up elsewhere. (Monopolies *do* operate like that. The newly divested AT&T had to learn quickly how to act in a competitive market.)

In the case of switched access, LEC access charges were averaged in to the IXC's per-minute rates. The FCC did not permit the IXCs to explicitly charge these back to their retail subscribers; AT&T and MCI, for instance, had to charge the same thing for a 1,000-mile call to a rural telephone coop with a 25-cent per-minute terminating rate as it charged for a 1,000-mile call to an RBOC with an 8-cent per-minute rate. But this *rate averaging* was not applied to leased-line services. So the IXC's price for an interLATA circuit would consist of three explicit components, the IXC's own PoP-to-PoP charge and the two LEC charges for each side. The LEC rates for this *special access* service were thus exposed to the actual bill payers.

And what rates these were! For some years before divestiture, AT&T's rates had been regulated based on the notion that it was entitled to a fair rate of return, no more, no less [1]. Just how that return was calculated was always controversial, of course, but by the late 1970s, as competition was taking hold, the FCC had adopted a complex set of *fully distributed cost* methodologies. The RBOCs, as monopolies, were originally subject to rate-of-return regulation. Typical approved rates were in the 10% to 15% range. At the federal level, these rates were supposed to apply to every type of service offered, not merely in the aggregate. States were expected to take greater liberties in allowing cross-subsidization of unprofitable residential and rural services by profitable business and urban customers.

Digital "high-capacity" services, so-called T1 lines, were only introduced to the general marketplace after divestiture. AT&T's old backbone network was primarily analog, and fiber optics were still just being installed on many routes, so T1 service really only took off after divestiture. The RBOCs had to set prices for them, which needed regulatory approval. Although rates for purely intrastate services were approved by state regulatory commissions, the intraLATA legs of interstate circuits were classified as interstate special access, subject to FCC jurisdiction.

The RBOC rates for high-capacity special access were high. They were unlikely to say it in public, but it seemed as if special access rates were being padded to make up for the loss of switched access revenue that could occur if much traffic moved off of the switched network onto leased lines. Corporations routinely built voice tie-line networks to connect their facilities together. So RBOCs might have wanted to extract their pound of flesh one way or another. After some investigation, the FCC found, for instance, that NYNEX had set its rates to earn a return of well over 50%, several times the allowable level. But the FCC did not step in to lower these levels. The way was being paved for competition, with excessively high rates acting as a kind of price umbrella for competitors to live beneath: Had the RBOCs been held to a fair rate of return, then the course of history might have gone differently.

States Supported RBOC Monopolies More Than the FCC Did

During the 1980s, most state regulators were rather friendly with the RBOCs. These newly minted offspring of the old Bell System began life with a considerable degree of goodwill. Known by the cutesy nickname "Baby Bells" and subjected to considerable restriction on their behavior, the RBOCs positioned themselves as the defenders of cheap home telephone service. And lacking both *direct* long distance revenue (never mind the huge cut they got via access charges) and the embedded base of terminal equipment rentals (which had gone to AT&T), they had to get their "universal service" money from somewhere. Regulators also understood that home telephone subscribers had more votes than corporate telecom managers.

This resulted in the continuation of a tradition of noncost-based pricing. Indeed some states adopted so-called "residual pricing." Under that scheme, the explicit goal of regulators was to minimize the "1FR" single-party flat-rate residential monthly base rate by applying profit-maximization monopoly pricing (sometimes called "incremental willingness to pay") to other services. The 1FR rate would thus be limited to the residual revenue requirement. This approach was not geared to produce economic efficiency nor did it later transition well into a competitive market. It did mean that many intrastate leased line rates, as well as long distance rates, produced rates of return that far exceeded the overall norm.

The FCC Preferred a Market-Based Approach

The FCC let high special access rates remain in place, presuming instead that competitive forces should be allowed to operate. The only problem was that there were not a lot of competitors out there. A few large corporations had built private microwave radio systems or fiber-optic networks. Westinghouse Corp., for instance, had built a large microwave network around the Pittsburgh area in the 1970s, and Digital Equipment Corp. built a few hundred miles of fiber-optic routes around Massachusetts and New Hampshire in the 1980s. These could interconnect directly with IXCs, *bypassing* the RBOCs. But almost everyone else had to pay full price.

The newly minted RBOCs did not expect competition to become a serious problem. They preferred to risk losing a few customers around the edges versus having to lower prices in general. Thus a market opportunity was born. Because access services were under FCC, not state, jurisdiction and thus did not require state authorization, new local service competitors were dubbed CAPs.

Satellite Bypass Did Not Quite Fly

One other avenue of bypass had already been created. In the early 1970s, the FCC decided that domestic satellites should be competitive—the "Open Skies" policy—in contrast to the then-monopoly Bell System. Although early satellite systems required large Earth stations that were operated by common carriers, one company pioneered the very small aperture terminal (VSAT), a satellite Earth station that could be located at the end-user premise, bypassing the usual telephone company-provided last mile. Satellite Business Systems (SBS) was set up as a joint venture between IBM, Aetna Life and Casualty, and Comsat General. IBM may well have designed its SNA, for years the backbone of its data communications strategy, as a way to work with satellites; its predecessor protocol Bisync was overly sensitive to delays. SBS service began in 1981. But it did not exactly set the world on fire.

Satellite service is crippled by the laws of physics. Geosynchronous satellites sit 22,240 miles above the equator. Radio waves travel no faster than 186,000 miles per second. (It's not just a good idea, *it's the law!*) The resultant delay is annoying for some data applications, and even more annoying to voice conversations. Satellites are thus primarily used for one-way broadcasting, for accessing very remote areas such as islands, and for specialized data applications that can tolerate the delay, such as point-of-sale systems. SBS was eventually bought by MCI, adding a few more satellites to its constellation.

Teleport Cracks the NYNEX Monopoly

The first competitive access provider grew out of the Teleport, an office facility in Staten Island, N.Y., with its own cutesy name no doubt inspired by the likes of Star Trek, where Merrill Lynch built a data center near a satellite ground station. Created in conjunction with partner Western Union, the Teleport Communications Group (TCG) also pulled fiber optics into Manhattan and Brooklyn, providing an alternative to NYNEX. Merrill Lynch was essentially bypassing NYNEX and taking in revenue from its neighbors at the same time, while laying the groundwork for a future competitive local telecommunications industry. The Wall Street area of Manhattan was, of course, one of the densest concentrations of bandwidth utilization in the world. If any low-hanging fruit was ripe for plucking, this was it.

But Merrill Lynch was not really in the telecom business, and Western Union was in severe decline, so a majority interest in TCG was acquired by several cable companies, including Cox, Comcast, Continental, and Tele-Communications Inc. (TCI). This provided an interesting avenue for expansion: The cable companies were already pulling fiber optics on their widespread rights-of-way, so providing high-capacity service to business via TCG was a good way to increase the value of their networks and perhaps help finance upgrades.

Competitors Outrace RBOCs to Provide Local Fiber-Optic Connections

TCG was the first CAP, but there were soon others. MFS was founded in 1988 by Peter Kiewit Sons', the large Omaha-based construction firm. It not only pulled fiber in a number of top-tier cities, it also took an early role in the growth of the commercial Internet. Its fiber rings in the Washington and San Jose areas were used by backbone ISPs, and in 1994 the company created the Metropolitan Area Ethernet–East and West (MAE-East and MAE-West) peering points, which remain major intersections on the worldwide Internet. But the major capital investment was in the local fiber networks. By 1996, MFS had pulled 3,183 miles of fiber-optic cable and had "lit" more than 5,700 buildings in 52 metropolitan areas [2]. This was no small investment. Investors had to expect a bright future to absorb such costs.

Another CAP, Hyperion Telecommunications, was created in 1991 by Adelphia Communications, a large cable operator. Hyperion's service area was a bit different: Rather than address the top-tier cities, it focused on Adelphia's footprint and on smaller markets where other CAPs were unlikely to go. For example, it built a route the length of the state of Vermont, then westward to Buffalo, N.Y., and served cities such as Erie, Pa., and Columbia, S.C. It later changed its name to Adelphia Business Solutions. (After bankruptcy, the assets were operated as Telcove.) Brooks Fiber Properties, a St. Louis-based CAP founded in 1993, also focused on second-tier markets, such as Providence, R.I., Little Rock, Ark., and Fresno, Calif. It went public in 1996. Of course none of these companies was to remain independent, or under its original ownership, much longer.

Metromedia Fiber Network (MFN, or sometimes known as MMFN) was founded as National Fiber Network in New York in 1993. Metromedia, the corporate umbrella of tycoon John Kluge, bought it in 1997. Its plan was to lay five million miles of fiber and provide dark fiber to bandwidth-hungry customers, mostly other carriers. But its expansion out of its home market did not really begin until after the Telecom Act was passed in 1996. Its boom-era mentality was explained in this 1999 quote. "Howard Finkelstein, MMFN's president, explains that a DS3 from the telco costs $3,000 a month, and a comparable MMFN fiber costs about $5,000. But the fiber can be lit at OC-12—14 times faster than DS3—for $500 more per month (assuming 10-year depreciation). This works out to about $400 per DS3 per month" [3]. True enough—undiscounted special access DS3 rates have remained at that level, but how many subscriber locations really need more than one? Most large Web hosting operations are at collocation centers, not city office buildings. This may look obvious now, but in 1999, investors simply expected demand *every-where* to grow to dark-fiber levels. And by 1999, the Telecom Act's unbundling requirement finally made ILEC dark fiber available to CLECs at much lower rates.

These were not the only CAPs, of course; they were just among the largest. After the Telecom Act, many more CAP/CLECs popped up. But after the Telecom Act, they were not usually called CAPs—access was only part of the local service package that CLECs were able to offer. Laying fiber in the ground was just the most expensive way to become a CLEC.

Overbuilding Each Other in Top Markets

As the boom took hold in the mid-1990s, and full-scale local telecommunications competition loomed on the horizon, CAPs sought to position themselves by being in all of the top locations. This led to an expensive mess, as the carriers followed each other in cutting trenches in streets in top cities around the country. A CAP's network had a backbone route, which typically ran down major city streets and then out to suburban office parks. But it then needed *lateral* routes into buildings, many of which were not quite along the way. This added to the amount of trenching needed. During the boom years, as the CAPs spent other people's money at a prodigious rate, many cities developed regulations to limit the disruption caused by all of this fiber installation. Sometimes this amounted to requiring the CAP with the trenching permit to allow others to share its trench, so that the street would be less likely to need further digging.

But was this worthwhile? It seems likely that the CAPs each told their financiers that they would take a sizeable market share away from the Big Bad Bells, whose overpriced access and data services were a drain on businesses. But it seems just as likely that the CAPs did not count on each other's dividing the take. Furthermore, the Bells began to request pricing flexibility. The FCC and some states allowed them to lower prices in hotly competitive markets. This slowed down the attrition of the RBOCs' biggest customers and resulted in even less business for the CAPs. So the CAPs were far from profitable. Fortunately for them, the boom was underway, and profit was simply not considered important. Special access tariffs in general did not fall, though—they remained the "sucker rate" for subscribers who had no alternative.

The Telecom Act Opens Local Service Competition

The Telecommunications Act of 1996 was passed on February 1 and was rapidly signed by President Clinton. A compromise reached after lengthy wrangling, all sides rapidly embraced it. This was widely viewed as the "big bang" that would reshape telecommunications. "I believe this bill means American companies will dominate the field of global telecommunications," said Rep. Jack Fields, chairman of the House Commerce Subcommittee on Telecommunications and Finances [4]. The Act made many changes in Title 47 of the U.S. Code, which had originally been enacted as the Communications Act of 1934. Some of these changes concerned broadcasting: The number of stations that a single company could own was greatly increased, leading to tremendous consolidation of broadcast radio. The Communications Decency Act would have criminalized Internet transmission of material "indecent to minors" had it not been overturned by the Supreme Court. Most cable television prices were deregulated. And most television sets were required to be equipped with a "v-chip," a feature enabling parents to block reception of programming tagged with signals indicating various types of content.

The impact on local telephone service was profound. Prior to its passage, a few states had already legalized local telephone competition: TCG and MFS, for instance, were already offering local service to businesses in New York and Massachusetts. But the Act required states to give certification to any and all competitive local exchange carriers that met basic qualifications. It set broad guidelines for competition, requiring incumbent LECs to provide unbundled network elements (UNEs) at cost-based rates. Of course the Act required considerable work on the part of the FCC to implement the detailed rules. It gave very short deadlines, essentially putting the FCC into high gear for the year, as regulations were pounded out.

But the very compromise that made the Act's final passage happen turned out to be its greatest weakness. The Telecom Act was, and is, a masterpiece of self-contradiction. To achieve a compromise acceptable to ILECs, IXCs, and potential CLECs alike, it contained many intentional ambiguities. The net result was a nearly permanent state of litigation, plenty of work for Washington's lawyers, but confusion for the businesses that had to operate within the law.

The Long Distance Business Declined

The Bell companies turned out to be the real winners. The Telecom Act required them to give up their local monopolies, but it also provided them with a means of setting aside the remaining restrictions on their behavior that had been present since their creation. Section 271 of the Act contained a 14-point checklist of obligations that the Bells were expected to live up to. If their state regulatory commission and the FCC agreed that they met the terms, then they would be given authority to sell interLATA services in that state. This process was not terribly easy, at least at first: The FCC in effect under President Clinton, then led by William Kennard, held the compliance bar high. The checklist required the Bells to unbundle numerous network elements, including loops, switching, transmission, and operator services. The Section 271 *process* often focused on the operational support systems (OSS)—hideously complex computer programs used for ordering and provisioning services [5]. The Bells had to open their OSSs to the CLECs, allowing UNE Platform and resale orders to be placed efficiently. However, once the Bush administration's FCC, led by Michael Powell, took over, state approvals fell like dominoes. And Powell did not believe that the Bells *really* had to unbundle switching; after all, CLECs *could* purchase switches of their own. So by 2004, the Bells were, in most cases, the second-largest retail long distance providers in their respective territories. To be sure, their customer base was mostly low-revenue subscribers who preferred one bill to two. But the Bells' gain in long distance market share was not nearly matched by losses in local service market share.

Between this loss of market share and the general decline in prices, the long distance business went into serious decline. The two-tier industry structure of the MFJ was being eroded. True, the Section 271 rules required the Bells to deal with their own long distance subsidiaries at something vaguely resembling arms' length; they still had to allow their subscribers to maintain separate interLATA and intraLATA primary interexchange carrier selection. But they were still really paying access charges to themselves, giving them an effective advantage over separate providers. The Telecom Act did allow the IXCs to become CLECs, of course; MCI and AT&T sold millions of lines of UNE Platform during the early 2000s, allowing them to keep some of their own access revenues. But that too fell victim to the changed regulatory environment.

CAPs Had Head Start on Both Service and Debt

The CAPs, of course, were the first CLECs. They quickly turned their "access" networks into "local" ones. A major change was their ability to get local telephone numbers. Interstate-jurisdiction access services included outgoing calling and 800 service; as CLECs, they could also get local prefix codes, blocks of 10,000 numbers assigned to a specific geographic rate center. CLECs also interconnected to the ILECs as peers. Access providers are "customers" of LECs, paying them for both origination and termination services [6]. CLECs, on the other hand, pay the ILECs to terminate calls that their subscribers originated but are paid by the ILECs when they terminated calls originated by ILEC subscribers.

This system, *reciprocal compensation*, turned out to be a major bone of contention when applied to Internet-bound calls. (See Chapter 9.)

Although the CAPs were the first CLECs to be up and running, they also had the biggest debt load. The Telecom Act made it possible to be a CLEC without spending the kind of money that the CAPs had spent. It authorized local competition via several avenues. *Total service resale* allowed the CLEC to sell ILEC-tariffed services under its own name, paying the ILEC's rate minus an "avoided cost discount" typically in the 15% to 25% range. Resellers required essentially no capital expenditure at all, save for billing systems. *Facilities-based* CLECs could provide their own networks in toto, interconnecting with the ILECs as peers, which is largely what the CAPs did, but they could also rent selected UNEs from the ILECs and thus reduce their expenditures. Once the Supreme Court resolved some ambiguities (at least temporarily) in a January 1999 decision [7], CLECs could rent *all* of the elements needed to provide telecommunications service, essentially using the same facilities as resellers but paying a cost-based rather than tariff-based rate. This became known as the UNE Platform and, over the next few years, largely replaced total service resale as the most popular means of providing competitive dial tone.

Investors, though, remained enthusiastic about the CAPs and CLECs for several years after the Act. CAPs were not profitable, after all, but compared with resellers they had "hard assets in the ground." In 1996, MFS purchased UUNET, the leading ISP, for about $2 billion. The following year, it was purchased by WorldCom for stock then worth more than $12 billion. To be sure, much of this price reflected rapid inflation in the value of ISPs, but the CAP/CLEC assets were being highly valued. In early 1998, WorldCom cemented its position as the leading CAP/CLEC by purchasing Brooks Fiber for about $3 billion. And also in early 1998, in a deal reminiscent of WorldCom's acquisition of MFS, AT&T acquired TCG, exchanging more than $11 billion in stock for the CAP, which the year before had purchased San Diego-based regional ISP CERFnet. This gave AT&T a local footprint in top markets and a credible ISP presence. But it did not give it profits. And in 1999, Bell Atlantic bought into its competitor, Metromedia Fiber Network. This was supposed to improve Bell's data services and out-of-region presence. Of course that did not turn out to be a very sound investment either. But at least Bell Atlantic, which shortly afterwards changed its name to Verizon, could easily afford the write-off that ensued.

Fixed Wireless as an Alternative to Fiber?

Wireless telephone services grew like Topsy during the 1990s, primarily for mobility. On the other hand, traditional telephone-company microwave systems were in serious decline. Before the coming of fiber optics, the bulk of the Bell System's long-haul backbone was analog microwave radio, but by the 1990s that network was almost entirely decommissioned. The surviving micro-wave manufacturers, however, were not sleeping on the job; high-capacity digi-tal microwave radios were available, with bit rates up to 155.52 Mbps (OC-3). Some vendors even came up with the marketing term "wireless fiber." Given the high cost of trenching fiber under urban streets, microwave radio could be an attractive substitute.

Two boom-era companies in particular, Winstar and Teligent, focused on developing this across the United States. The basic business model was to get an office building owner's permission to put an antenna (typically a small dish) on the building, then wire up the building's tenants. The companies would put their base stations on enough high points to provide coverage around a city. Because they were generally operating in the 24 to 40 GHz frequency bands, where radio waves were easily absorbed by rainfall, their reliable range was usu-ally only 2 to 3 miles, but this was adequate for urban applications.

The problem with the business model turned out to be time. After enter-ing a city, it took time to get many building owners' permission to put up the rooftop antenna and string new wire to the tenants. Many buildings had con-gested conduits, making the wiring difficult. CLECs that leased ILEC loops could build onto the existing infrastructure, without having to fish the walls or get the landlord's permission. So the time to revenue was rather long, often close to a year. Thus the companies needed plenty of capital to continue expansion. Like other facilities-based providers, these wireless CAP/CLECs were able to raise plenty of capital during the early boom years. Some of this came in the form of vendor financing. Winstar was a major customer of Lucent Technolo-gies, which reportedly lent Winstar about $700 million between 1996 and 2000 [8]. Several billion more dollars came from other sources; however, it was largely lost in the company's 2001 bankruptcy filing. Winstar's business was acquired in late 2001 by IDT, a New Jersey-based telecom company that had managed its cash much better during the boom years. The price was only $42.5 million, in cash and stock. IDT Chairman Howard Jonas said at the time of the acquisi-tion, "This is an incredible deal. It might not top the Dutch settlers buying the Island of Manhattan for $24, but it comes pretty close. With almost $5 billion in assets and about $200 million in annual revenue, Winstar has great potential" [9]. But IDT Solutions, as the subsidiary was renamed in 2003, remained in the red into 2004, with no profit in sight.

Teligent's original story featured its charismatic CEO, Alex Mandl, a former AT&T executive. It began with a trade of some valuable spectrum licenses that its founders had acquired earlier. But it also expanded too fast, run-ning out of cash in 2001. It filed for bankruptcy, emerging in 2002 under the ownership of its former creditors. Another CAP/CLEC, originally called

Nextlink but later XO Communications, also built fixed wireless systems in most major markets. Its network was more diverse, featuring a mix of CAP fiber, a backbone network, a large ISP (formerly Concentric Networks), and a multi-product CLEC. But, as noted in Chapter 5, it too went through bankruptcy.

A different approach to using wireless service to compete with ILEC fixed services is known as wireless local loop (WLL). This provides retail-level services to homes and smaller businesses using lower-cost radio systems. WLL has proven useful in some developing countries, where fixed loop facilities were not available. But it has been more talk than action in the developed world. Some of this, especially in the United States, may be the result of regulatory policies. Rural ILECs are given huge subsidies to provide wireline service at monthly rates even lower than their low-cost urban counterparts. The subsidy is tied to their level of investment, so there has been little incentive for them to substitute lower-cost WLL technologies. The United States also has no radio frequencies set aside for WLL. ILECs in the most rustic areas, for whom even subsidized copper loops would be impractical, are allowed to make use of land mobile and cellular telephone frequencies for a primitive analog WLL called basic exchange telecommunications radio service (BETRS). Most other countries, however, have a licensed fixed-wireless spectrum allocation at 3.6 GHz that is available for WLL.

Although several vendors have produced equipment for this purpose, overall take rates have been limited. Perhaps the best-known operator was the United Kingdom's Ionica, which gained 60,000 subscribers between 1995 and 1999, when it went into receivership. Ionica had developed a unique WLL technology, which was manufactured for it by Nortel. To be sure, its failure was not entirely the fault of the technology. The company had a high overhead (about a thousand employees by the time it failed) and was reported to be spending money in the style of the day, which is to say extravagantly, on corporate overhead. Some other WLL niche players have remained in business, even in the United States. Western Wireless adapts PCS technology for fixed users, serving a number of rural states; much to the chagrin of the ILECs, it has qualified for "eligible telecommunications carrier" status and thus can share in the subsidies granted to rural telephone companies, even though it not an incumbent.

Overexpansion Led to More Bankruptcies

A number of other companies jumped in to the CAP/CLEC business during the boom, only to spend beyond their means. One of the largest was McLeod USA, based in Cedar Rapids, Iowa. The company approached the CLEC business aggressively, taking a sizeable market share in several Midwestern states. It built an extensive fiber-optic network; although eventually national in scope, it was especially well developed in Iowa and even crossed the sparsely settled lands of Nebraska, Wyoming, and Idaho. For a time, the company even dabbled in the ILEC business, buying Illinois Consolidated Telephone, a rural carrier with more than 75,000 lines.

Analyst reports during the later boom years showed that McLeod USA was probably profitable in its oldest markets, the ones opened in 1996–1997. But its rapid pace of expansion caused its bottom line to remain negative, and when the bottom fell out of the market, there was no more money left to cover its losses. So it too joined the Chapter 11 club in January 2002. It downsized, sold Illinois Consolidated back to the original family that had owned it, and sold its directory business; the reorganized CLEC was finally EBITDA positive.

Another CLEC, e.Spire, had a CAP unit called ACSI Network Technologies, which also provided fiber construction services to other companies. It spent the boom years expanding in various cities around the country, but like many others, ran out of cash and met the bankruptcy judge in early 2001. St. Louis-based Xspedius, founded by a veteran of Brooks Fiber (and thus essentially funded by the high price WorldCom paid for Brooks during the boom years) picked up its assets the following year for $68 million, vaulting itself onto the national scene for a meltdown bargain price.

Hybrid Fiber-Coax (HFC) Gave Cable Providers an Advantage on "Triple Play"

One class of companies was able to build facilities in parallel to the incumbent telephone companies and make money at it. The cable television (CATV) industry had long coexisted comfortably alongside the telephone companies. For many years the two industries avoided challenging each other directly, even though it was long a possibility. The FCC recognized this by prohibiting local telephone companies from owning cable companies in their service areas, except for outlying rural areas where the cost of two networks would be prohibitive. Telephone companies were allowed to lease facilities to cable companies but not actually sell the services; the two industries rarely even entered into that level of cooperation, but neither did they compete.

In 1982, start-up Local Digital Distribution Corp. introduced two products designed to foster competition with the LECs. Radio packet controller (RAPAC) was a point-to-multipoint radio system designed for the new 10.5-GHz band [10]. Cable packet controller (CAPAC) was a cable modem. To be sure, the public Internet was a decade away, so the mass market was not there, but CAPAC could have been used for linking corporate sites, telecommuting, competitive access to long-haul data networks, or various other purposes. But the product was a market failure. A corporate executive explained the problem succinctly: Cable companies were not in the carrier business, they were in the entertainment business [11].

RBOCs Took the Threat Seriously

By 1985, the still-young RBOCs were aware of the potential threat from cable, even in the absence of overt acts by the cable industry. The ITU's Broadband ISDN program, best known as the source of ATM network technology, was begun, in part, as a response to that potential threat. The best defense being a good offense, the telephone industry set out to develop its own set of technologies that could carry video to the home. B-ISDN was to be the telephone companies' own triple play: voice, data, and video all on one network. It depended on pulling optical fiber to the home (FTTH) and was designed to carry high-definition television at an estimated required bandwidth of 100 to 120 Mbps. (This turned out to be far more than actually required.) So the people working on B-ISDN saw it as a very long-term goal, something that would probably not see widespread deployment until, say, the late 1990s!

A lot happened between 1985 and the late 1990s, of course! In the narrow context of B-ISDN plans, the first practical ATM switches did not hit the market until after 1990, when Fore Systems introduced an ATM-based 100-Mbps LAN switch. This tilted ATM development away from the metropolitan and wide area networks of B-ISDN and toward LANs, and the most important forum for ATM standardization moved from the ITU to the ATM Forum, which was founded in 1991. In the early 1990s, RBOCs in numerous states made promises that they would provide high-speed (50 Mbps or more) fiber to the home within the next decade, provided that they were freed from rate-of-return regulation. That resulted in the Bells' being moved to a new form of regulation, *price caps*, which allowed them to adjust their prices within a certain range (typically keyed to the rate of inflation, minus an annual expected-productivity-boost factor) and keep any remaining profits. The Bells profited handsomely from this but never pulled the fiber [12]. And Bell-company video-on-demand trials ended in the mid-1990s.

Hybrid Fiber-Coax Is Developed

Now let's set the wayback machine to 1989. Fiber optics had become ubiquitous in the long-haul and interoffice telephone networks, but the CATV industry was still using coaxial cable. Because coaxial cable has substantial attenuation that rises with frequency, CATV networks of the day required amplifiers every few blocks. Amplifiers added noise; adding two-way services required all of the amplifiers in the path to be upgraded. As many as 30 amplifiers could sit between a head end and the subscriber; bandwidth was typically limited to around 350 MHz, allowing perhaps 50 channels per cable. CableLabs, an organization sponsored by a number of multiple system operators (MSOs), commissioned the consulting firm Arthur D. Little Inc. (ADL) to study the role of fiber optics in the CATV industry. The ADL study determined that the most cost-effective architecture would use low-loss fiber optics (in analog mode, carrying the radio frequency spectrum modulated onto light) for the relatively long-haul trunking between the head end and the neighborhood, with coaxial cable, and no more than three amplifiers, for the final subscriber drops. This scheme came to be known as hybrid fiber-coax (HFC) and was adopted universally by the industry.

During the 1990s and beyond, virtually all new cable builds used HFC architecture, whereas older systems were upgraded by replacing the old coaxial trunks with optical fiber. This resulted in two-way networks with greatly increased bandwidth (550 to 890 MHz) and thus far more channel capacity. It also meant that the cable companies, not the ILECs, were the ones with fiber in the residential neighborhoods and the ones with high-speed two-way transmission into the most homes. Although the ILECs were promoting digital subscriber line services, those were very sensitive to loop length and quality, so roughly only half of American homes would be able to subscribe to DSL if every central office had it. HFC, on the other hand, provided uniform service throughout its coverage area, so every home in a served franchise could get all available services.

HFC can provide voice, data, and video services quite readily. In North America, where television channel 2 began at 54 MHz, frequencies below 42 MHz are generally available for upstream (reverse) transmission on the coax. (On the fiber, upstream generally uses a different strand from downstream. European systems can provide up to 65 MHz for upstream, there being little VHF television there.) However, frequencies below around 18 MHz are not suitable for high-speed use. This still leaves room for one or more upstream cable modem channels (either of the 6-MHz bandwidth of a TV channel or, in some cases, 3 MHz), as well as a channel for telephony. Several vendors created dedicated HFC telephony systems during the 1990s. AT&T Technologies, even before it became Lucent, created an analog system capable of mapping 240 telephones onto a TV channel; although piloted by SNET in Connecticut, it was discontinued before mass deployment. Tellabs and Arris Interactive, however, both developed successful TDM digital telephony systems and deployed millions of lines apiece on various operators' networks.

Cable Modems Sparked a Cable ISP Boomlet

When the Internet boom hit, many cable companies were thus in an attractive position. They had shiny new HFC networks, and the cable modem industry had been reborn to take advantage of them. Early cable modems were proprietary; the customer-side modem and the head end cable modem termination system (CMTS) had to come from the same vendor. In 1996, CableLabs again retained Arthur D. Little Inc., this time to put together a consensus standard for cable modems, called Data Over Cable System Interface Specification (DOCSIS). By 1998, DOCSIS-compliant cable modems and CMTSs were on the market, facilitating volume rollout. In 1999, DOCSIS 1.1 was specified, adding the ability to dynamically reserve bandwidth to individual modems, a feature especially useful for cable telephony. This led to another set of Cable-Labs standards, PacketCable, that standardized a method of providing Voice over IP over DOSCIS 1.1 cable modems. Hardware prices were set to fall as volumes rose.

By 1995, pioneering MSOs such as Continental Cablevision had installed cable modem networks in a few cities. Continental had built its own in-house ISP (Continental Express), which remained intact after its acquisition by US West, which renamed the company MediaOne (and its ISP Highway One). But the largest MSO of the early 1990s, TCI, was not so comfortable handling this newfangled Internet thingie by itself. It did not dawn on the leaders of the cable industry to work with established ISPs (not that many were well established by 1995) or, for that matter, to set themselves up as Internet access service providers selling wholesale access to other ISPs [13]. So along with venture capital firm Kleiner Perkins, TCI created @Home, an ISP dedicated to the cable industry. @Home signed multiyear exclusive deals with TCI and several other MSOs, some of whom took equity positions, and became one of the largest ISPs in the country; it also went public, raising billions of boom bucks. @Home might have been a success story, but it too fell victim to boom insanity. In 1999, all things Web were hot properties, especially "portals," Web sites that pointed to other Web sites and provided Web search services, liberally brushed with advertising. @Home offered a measly $6.7 billion for the number two portal, Excite. The merged company, Excite@Home, sold for nearly $100 a share at its 1999 peak. But in 2001, it joined the parade at the courthouse. As part of the bankruptcy deal, AT&T, which had purchased TCI a few years earlier, got ownership of the ISP assets while the portal's assets were auctioned off for a pittance. AT&T phased out @Home before selling its entire cable operation to Comcast.

Fiber to the Home Kept Moving Further into the Future

The cable ISP stock business may have been weak, but the cable Internet business itself prospered, outselling ADSL by a substantial margin. In the key residential and small business sector, FCC statistics showed that in December 1999, there were 1.4 million "high-speed" (more than 200 Kbps in at least one direction) cable modems, versus 291,000 ADSL lines, and only about a thousand fiber-optic (FTTH) lines nationwide. By December 2003, ADSL had grown to 8.9 million lines, whereas there were more than 16.4 million cable modems. ADSL's market share was rising, from 16.3% to 34.4%, but cable retained 63.2% of the broadband market [14]. And whither FTTH? By the end of 2003, not quite 20,000 homes were connected nationwide. Although regulators, telephone companies, and especially equipment makers fretted about the failure of FTTH, they were perhaps blindsided by the success of HFC, which delivered the bulk of the benefits at far lower cost.

Some sectors of the industry, especially fiber-optic and equipment manufacturers, may have genuinely fretted about the failure of FTTH to catch on. Predictions made in the 1980s had it widespread by the turn of the century, but that did not happen. HFC cable was one reason; ADSL was another. And very-high-speed DSL (VDSL) was on the horizon, already catching on in Asia (where local loops were shorter). So regulatory efforts to encourage FTTH were, in a sense, a form of "industrial policy" that violated the free-market principles that these same regulators ostensibly held dear. FTTH was a technological bridge too far, a leap that could not be cost justified when less-costly, more evolutionary technologies kept stretching the capabilities of both twisted-pair and cable plant. Although it was almost undeniable that FTTH would *eventually* happen, betting on its widespread success proved to be a bad idea. Regulatory efforts to promote it were perhaps akin to a government program to ensure that your dog really did get enough cheese. Even if you did not have a dog [15].

Cable companies' success in serving residential markets did not translate to a significant share of the business market; for that, the traditional CAPs remained the primary source of alternative broadband facilities. This was largely because cable systems did not even go into many business areas. Cable was designed for home entertainment, not business. And the bandwidth characteristics of HFC were not optimized for business use. Fiber to the home was not necessary, but fiber loops were a natural for business locations.

A single mid-sized business PBX may require several DS1s of trunk facilities. A typical cable telephony system does not have such highly concentrated capacity. And cable modems are generally highly asymmetric, with far more downstream bandwidth than upstream. That corresponds to residential data demand but not business use, which is far more symmetric. Voice bandwidth, of course, is also symmetric; its upstream requirement is a limiting factor for PacketCable capacity on a given piece of coax. DOCSIS 2.0, first marketed in 2003, dramatically improves upstream capacity (raising the maximum rate from 10 Mbps to 30 Mbps), so there is more future likelihood of business use of cable networks.

Exuberant Prices for Existing Systems as Industry Consolidated

The boom years had their own impact on the cable industry. As with so many other aspects of the communications and technology industries, stock prices rose, often to ridiculous levels, which seemingly encouraged mergers and acquisitions. Between 1993 and 2003, the bulk of the industry consolidated into a few large players, while fortunes were made and lost.

To be sure, the cable industry had long been a game of Monopoly of sorts. MSOs routinely bought, sold, and traded territories, often trying to put together larger, more contiguous areas that could be operated more efficiently. This was a natural outcome from the way that many original cable franchises were awarded. In some states, franchises were awarded on the city or town level. In some large cities, such as Los Angeles, city governments encouraged rapid buildout by awarding multiple franchises. Several operators then built head ends and started building out from there; when they ran into a street that another operator had already wired, they stopped. Contrary to common belief, most franchises were nonexclusive; cable operators simply found it a better investment to be the first to serve a given location, not the second. Thus the monopolies that resulted were very much of the "natural" variety.

By the early 1990s, cable service had moved out of its rural roots and become widespread across the country. New cable-only channels helped make cable popular in urban areas that had good television reception. As cable penetration increased, so did system valuation. The promise of cable telephony and cable data services was also beginning to look quite real. HFC's promise was being validated by several large telephone companies, including GTE, Southern New England Telephone, Pacific Bell, and Ameritech, which began to build their own HFC networks. In the mid-1990s, they found cable telephony to be a bit harder than expected—the equipment of the day was primitive and powering customer-premise telephony adapters was proving troublesome. But those four large ILECs did not abandon HFC until they were acquired by HFC-averse Bell Atlantic and SBC.

Bell Atlantic was not always so wary of cable. In 1993, it offered $33 million for TCI, the largest MSO with 9.7 million subscribers. The deal collapsed, though, in the face of regulatory difficulties. TCI's network was among the most primitive—the company had been cobbled together from a long string of acquisitions, and its chairman, John Malone, was known for his interest in programming, not system maintenance. Some wags suggested that Bell Atlantic's CEO, Ray Smith, an amateur thespian himself, had wanted to bask in the glory of a Hollywood-centric industry.

But one RBOC moved into cable in a big way. Denver-based US West first got its feet wet by acquiring two cable companies in Atlanta, safely distant from its telephony service area, in 1994. The 446,000 subscriber acquisition cost about $1.2 billion, about $2,700 per subscriber. The company's cable unit adopted the name MediaOne. It also paid $2.5 billion for a 25% interest in Time Warner Entertainment, the Time Warner unit that included a large MSO. In 1996, shortly after the Telecom Act passed, it dove in head first, buying Continental Cablevision, with about four million subscribers, for almost $11 billion. (Continental's systems in US West territory, such as its Minneapolis, system, had to be sold off as part of the deal. Charter Communications, for instance, paid almost $2,100 per subscriber for Minneapolis.) Industry consolidation included many smaller acquisitions, usually at much lower per-subscriber prices; rural systems with fewer than 10,000 lines typically sold for $1,000 to 1,500 per line [16].

Given Continental's already HFC-rich network, it seemed likely that US West would be interested in using the network to provide competitive telephone service. Indeed this did happen; MediaOne became a major CLEC, using TDM digital cable telephony systems from Tellabs and Arris. Cox and others also added telephony to some of their cable systems, though rollouts were far from universal.

In 1998, AT&T made a deal worth approximately $48 million to acquire TCI, more than $3,000 per subscriber. This propelled the telephone giant to first place in the cable business but left it with a huge job to do because TCI's network was largely viewed as a "fixer-upper." John Malone was left in charge of a separate programming unit, Liberty Media, which was eventually spun off. In 1999, US West put MediaOne up for sale. Comcast's offer was followed by AT&T's offer of $23 million, worth $4,632 per subscriber. This was the height of the cable industry's irrational exuberance phase. AT&T Broadband had cemented its leadership in size, but the parent company's debt headache was growing worse. How could even triple play cover such acquisition costs? In 1996, the average cable system sold for $2,115 per subscriber; AT&T helped propel that to $3,819 in early 1999. The average acquisition's cash flow multiple went from 11.4 to 16.3 in that time [17]. In 2002, AT&T sold its cable unit, which by then had about 22 million subscribers, to Comcast, in a deal worth about $72 billion, under $3,300 per subscriber. AT&T had lost tens of millions of dollars in its foray into cable. Comcast became the undisputed king of the hill, requiring an FCC rule change to be allowed to have such a large share of nationwide cable subscribers. Not only that, but by 2002, the bulk of network upgrades had been completed. Comcast had grown from being a modest-sized regional provider in the early 1990s to being one of the country's top media conglomerates a decade later. It did this, in large part, by exercising what turned out to be a rare skill, knowing when *not* to make an acquisition in an overheated market.

Overbuilding Was Often a Costly Disaster

The Cable Act of 1992 prohibited exclusive franchises. Local franchising authorities, such as cities and counties, were allowed to impose certain obligations on franchised cable providers, but they had to allow new companies to apply for franchises, even if one was already in place. This did not happen very often for the simple reason that the economics of *overbuilding* usually stank. The cost of building a cable network is largely based on the number of plant miles required. The cost *per home passed* is thus a function of residential density. But the cost *per subscriber* is a function of the cost per home passed and the take rate. If an incumbent cable operator serves 60% of homes passed and an overbuilder takes one third of its business away, it will still have only half as many subscribers per mile as the incumbent.

But given the ridiculous prices that were being paid for cable systems during the boom, overbuilding was tempting. A new HFC network could be built, in the late 1990s, for perhaps $20,000 to $30,000 per mile, in a suburban area with aerial (not underground) utilities. So if an overbuilder signed up 10 homes per mile, it was coming out ahead of, say, AT&T. But that did not translate to profits. A handful of companies did invest heavily in overbuilds; for the most part, they lost money.

One of them, West Point, Ga.-based Knology, was a spin-off of ITC Deltacom, originally the small rural ILEC called the Interstate and Valley Telephone Co. (ITC). ITC was also the original parent company of ISP Mindspring, which was merged into Earthlink in 1999. Knology overbuilt cable networks in a number of cities, mostly in the South, and acquired other systems, including some lines originally pulled by GTE. But it remained unprofitable a decade after its founding; it had, however, achieved EBITDA profitability, putting it ahead of many boom-era companies.

The other big overbuilder was New Jersey-based RCN Corp. The company emerged in 1997 out of a set of predecessor companies, some of which were connected to Kiewit's MFS. RCN's role was essentially to take over the residential side of the business ("residential communications network"). Its largest owners ended up being fellow Kiewit spin-off Level 3 and Microsoft co-founder Paul Allen's Vulcan Ventures (which also took a controlling interest in Charter Cable). Allen is both famously rich and a famously feckless investor, having lost money on an uncountable number of companies (which Allen, founder of Seattle's Science Fiction Museum, might say were simply ahead of their time). RCN was one of his biggest losers. Before declaring bankruptcy in 2004, it plowed through money at a prodigious rate, losing close to $4 billion in the previous five years.

But to some extent RCN's problems were a result of its boom mentality, not inherent in all overbuilds. For example, when it first entered the Boston market in the late 1990s, it spent heavily on television and side-of-buses advertising, among other media. Yet its services were only available at the time in a tiny fraction of the metropolitan area. The best it could do for most potential customers was to resell ILEC telephone service and provide dial-up Internet service via a series of ISPs that it had acquired during the boom, at boom prices. It did eventually overbuild much of Boston and nearby cities and towns, but it was overbuilding some of MediaOne's, which became AT&T's, which became some of Comcast's better cable plant. RCN was expanding into numerous major markets that had a large number of homes passed, but it was not apparently focused on the bottom line.

Overbuilding is not a guaranteed failure. Under proper circumstances, especially on a small scale, an overbuilder can achieve some success. And costs have fallen substantially since the boom, as the supply and demand equilibrium has changed for both materials and labor. Many current overbuilders are municipally owned electric utilities that add cable or even fiber to the home to their product mix. They are often motivated by economic development goals—superior telecommunications, they suggest, can help attract industry and bring other benefits to the population. This often works best in small towns that have been neglected by large incumbents. But overbuilding cannot be counted on as a general approach to bring down telecommunications rates. Whether aimed at residents and leading with video programming, or aimed at business and leading with fiber, overbuilders have a high burden to overcome the natural monopoly benefits that accrue to incumbents.

Endnotes

[1]   Rate of return is calculated in terms of profit as a percentage of the total *rate base*, the undepreciated value of the company's cumulative investment.

[2]   Garfinkel, S., "Where Streams Converge," *HotWired*, Sept. 11, 1996, http://hotwired.wired.com/packet/garfinkel/97/22/index2a.html.

[3]   Isenberg, D., "LEC of the Future Has Arrived," *America's Network*, Dec. 1, 1999, reprinting Isenberg's *Smart Letter No. 31*.

[4]   Scherr, E., "Fact Sheet: Telecommunications Act of 1996," *U.S. Information Agency*, 1996.

[5]   The Bells were major users of Telcordia's OSS, known as TIRKS (originally Trunk Integrated Record Keeping System but widely expanded in capability since its name was given), which was rumored to be one of the largest computer applications in production in the world, with more than 15 million lines of code.

[6]   The originating leg of a switched access call, when a local subscriber initiates an interLATA call, is billed to the IXC; as such, originating access is essentially billed as a collect call to the IXC.

[7]   *AT&T Corp. et al. v. Iowa Utilities Board*, Sup. Ct. 97-826 (1999).

[8] "Winstar Files for Bankruptcy, Blames Lucent," *ISP Planet,* April 19, 2001, http://www.isp-planet.com/cplanet/news/000104/april19winstar.html.[

[9] IDT/Winstar press release, Dec. 20, 2001. http://www.iat.net/corporate/press/release/24b_asp.

[10] In 1980, Xerox Corp. successfully petitioned the FCC to make a band of frequencies at 10.5 GHz available for a competitive common carrier digital radio service that it intended to market as XTEN. The frequencies were allocated but Xerox had second thoughts. Eventually the frequencies were reassigned.

[11] Presentation given by Charles Christ of Local Digital Distribution Corp. at the *National Conference on Local Networks,* Washington D.C., 1982.

[12] A number of articles about this have been published by Bruce Kushnick and his nonprofit organization Teletruth (http://www.teletruth.org) has detailed alleged overcharges resulting from broken promises to install broadband. For example, Teletruth's home page in mid-2004 asked, "Broadband Fraud in PA? Are Customers Owed $1,135.00 from Verizon for a Fiber Optic Network They Never Received?"

[13] The FCC agreed with the MSOs that cable networks were not common carriers and were thus not obligated to provide "open access." A few smaller cable operators did, however, choose to work with established local ISPs, and after the @Home exclusivity contracts ended, major operators such as Comcast did allow a few large ISPs to share their networks.

[14] "High Speed Services for Internet Access: Status as of December 31, 2003," FCC Wireline Competition Bureau, Industry Analysis and Technology Division, June 2004.

[15] This alludes to a dog food commercial that began by asking the viewer, "Is your dog getting enough cheese?"

[16] Cable transactions are tracked by the FCC and reported by the Media Bureau. http://www.fcc.gov/Bureau/cable.

[17] *Sixth Annual Report on Competition in Video Markets,* FCC, 99-418, January 2000.

[18] *Annual Assessment of the Status of Competition in Markets for the Delivery of Video Programming, Fourth Annual Report,* FCC, 97-423, December 1997.

8

DLECs and ELECs: An Exercise in Oversupply

Perhaps the most important real innovation that the telecom industry brought to market during the boom years was DSL service. DSL, broadly construed, is a family of technologies that extend the capabilities of copper local loops to deliver high-speed data service, bypassing the switched telephone network. The introduction of DSL provided the telecom industry with a way to compete with cable modems, without requiring large-scale capital investment to replace the local loop with optical fiber. It also nearly coincided with the passage of the Telecom Act, which led to a highly competitive market.

For the sake of clarification, it should be noted that there are several different DSL technologies, which tend to address different market requirements. Asymetric Digital Subscriber Line (ADSL) is the mass-market offering. It provides downstream (CO-to-subscriber) data speeds of up to 6 Mbps on the best short local loops, with maximum speed declining rapidly with distance; its upstream speed can be as high as 640 Kbps, though much lower rates are more common. ADSL's key feature is that it can share the local loop with an analog phone line. HDSL is the modern substitute for T1 transmission, providing 1.544 Mbps on a 12-kilofoot loop [1]. HDSL is extensively used inside ILEC networks, primarily for T1 service delivery, but it rarely sold as a DSL service per se. Symmetric DSL (SDSL) provides two-way capabilities at maximum speeds that again vary with the loop length, typically a bit over 2 Mbps on short loops but falling to the 300-Kbps range on very long loops. VDSL potentially delivers up to 50 Mbps over short distances, such as a "fiber-to-the-curb" or neighborhood-node scenario. ISDN DSL (IDSL) delivers about 144 Kbps by using unswitched ISDN basic-rate loop technology. That is not exhaustive;

there are other, lesser-known technologies that sometimes share the "DSL" moniker, as well as ongoing upgrades such as ADSL2+.

DSL First Failed as a Video Offering

The first deployments of ADSL, in 1993, were trials that predated the consumer Internet market. The key application was video on demand (VOD). ADSL can reach T1 downstream speeds (1.5 Mbps) over loops of up to 12 kilofeet, which, according to industry practice [2], is the maximum length for new copper loop deployments. Video can be compressed to 1.5 Mbps using the MPEG-1 standard, which is sometimes called "VHS quality"; this technology was readily available in 1993. Upstream bandwidth demand was very small, since it was mainly used for "remote control" functions. But other aspects of video on demand were not ready. In one VOD trial, the telephone company was rumored to have set up a roomful of videocasette players, with an operator to fetch the requested tape and patch it into the subscriber's line! Within a short time, ADSL was consigned to the back burner.

The Telecom Act Invites Novel Use of Unbundled Loops

By the time the Telecom Act passed in 1996, the Internet boom was beginning, and demand for consumer access at higher than dial-up speeds was intensifying. ISDN was the medium of choice for many users, but the incumbent carriers were not entirely happy about it; ISDN was still dial-up, and unless carriers could extract per-minute charges from users, they did not want to deal with the relatively high usage that characterized Internet subscribers. The market was finally ready for ADSL.

Although there were some early rollouts of data-oriented ADSL by some of the ILECs, the Telecom Act opened the door to new CLECs. It required the ILECs to make the local loop available as a UNE at rates based on "forward-looking cost." Even before the FCC, the courts, and the states could collectively decide on what these rates would be, entrepreneurs were setting up CLECs. Although the incumbent LECs, as monopolies, always attempted to be "all things to all people," CLECs could be specialists, and DSL was an inviting specialty.

Collocation Did Not Come Cheap

Covad Communications, the pioneer at competitive DSL, was founded in 1996 to be a data-oriented CLEC, a specialty sometimes nicknamed a "DLEC." Based in San Jose and founded by Intel veterans, it did not set out to offer any kind of "dial tone" service or, for that matter, make use of the switched telephone network at all. As a CLEC, it gained the right to rent unbundled loops, to collocate its equipment in the central offices, and to lease (at UNE rates usually far below special access tariffs) unbundled bandwidth between central offices in order to interconnect its network [3].

The rules that Covad worked under in 1996 and 1997 largely defined its business. Although ILECs had to allow collocation, they initially required every collocator to erect a wireframe cage in the central office, usually of 100 square feet. By the time all the fees were paid, the nonrecurring charge for the cage was often on the order of $100,000! Atop this, the CLEC had to pay monthly rent on the floor space, plus the ILEC's fees for dc power. This type of investment only made sense in the largest central offices, where there was some chance of being able to amortize it among a large enough customer base.

Another rule in effect at the time required the DLEC to pay the entire cost of the local loop. ADSL was designed for line sharing: ILEC consumer ADSL services made use of the high-frequency spectrum (the frequencies about 20 kHz) on a loop whose baseband (audio frequency) spectrum was already in use for analog telephony. So the ILEC had no incremental loop cost for ADSL. Covad did not provide the baseband telephone service, so paying for the entire local loop would have made it uncompetitive in the ADSL marketplace. (Loop costs varied city by city, most ranging between $10 and $20 per month.) Instead, it focused on providing SDSL to business subscribers, which were willing to pay more per month. Working in conjunction with many independent ISPs as its wholesale customers, Covad positioned its SDSL service as a lower-cost alternative to ILEC T1 channels, which frequently cost an ISP more than $200, even within a city.

It was hard to see where Covad or for that matter any DLEC was going to make its profits under those conditions, but if it did pick up a substantial share of business data lines, and only built out in large enough central offices, then it had a chance of making a profit. Early DLECs had to pay several hundred dollars per line for their DSLAMs (the central office end of DSL), plus monthly backhaul fees for every collocation—to connect their DSLAMs to their regional hubs—of several hundred to more than a thousand dollars per month. But this was the late 1990s, a time when profit was largely seen as an obsolete concept; DLECs were building revenue, riding the wave, positioning themselves, building a "story," whatever!

ILECs Controlled the Mass Market for DSL

The ILECs began to roll out ADSL in 1996, but their pace was leisurely. Deployment focused on the largest central offices, which had the fastest path to profitability but also seemed to follow the cable companies. If a cable company was installing cable modem service in a given area, ADSL seemed more likely to be deployed. In most cases, cable outsold ADSL, but ADSL was a far higher-volume product than the business SDSL that the DLECs featured.

In 1998, the FCC issued a ruling that broadened the data CLECs' horizons. It created a new UNE, referring to the data portion of a shared loop as the "high frequency element" (HFE), or "high frequency portion of the loop" (HFPL), ordering ILECs to make it available to CLECs at the same price they imputed to their own ADSL services. A few ILECs had previously made shared loops available at a discount to the full loop rate; Covad, for instance, had been paying approximately $5 per month to US West. But the new rule hit the ILECs hard. Most had imputed zero cost to their own shared-loop ADSL to make it more competitive with cable, so that became the monthly fee for CLECs using the HFE. (Sometimes a small administrative or testing fee could be added, but even that was often under a dollar.) CLECs were now able to compete for the consumer ADSL business.

Another aspect of the FCC's ruling impacted collocation. The requirement for cages was removed. Cageless collocation became available at a much lower price. A CLEC could now collocate a single rack; the typical nonrecurring cost for a single rack was around $10,000 and the monthly rent was under $100. (Power, however, could still cost several hundred dollars per month.) The early CLECs had already filled many of their collocation rooms with often nearly empty cages; cageless collocation not only lowered the price of admission, but it made space available for more CLECs than before.

Capital Poisoning Led DLECs to Overexpand

Covad had received a good reception from capital markets. Receiving ample venture funding, it went public in January 1999. But Covad was far from alone. Several other companies also received extensive funding to open DSL-oriented CLECs, and they were in a hurry to spend it. Northpoint, established in San Francisco in 1997, raced Covad to have coverage of the same prime markets. It also went public in 1999. And Colorado-based Rhythms NetConnections did the same, having both been founded and gone public in the same years as Northpoint. Both Covad-wannabees had ample capital from the best venture firms, which seemed to be impressed with their name-brand management. Northpoint was led by Liz Fetter, a Pacific Bell veteran, and Michael Malaga, previously with MFS WorldCom. Rhythms was led by Catherine Hapka, a well-placed former US West executive.

And in the boom years, profits were not important; Northpoint's Malaga was quoted in 1999 as pooh-poohing a $100 million loss with a line that could be a mantra from the era: "To be big you have to spend big"[4]. And big spending was the rule, as all three DLECs ran huge "burn rates." But burn rates did not lead to customers. According to a Yankee Group report, as of mid-1999, Covad had 31,000 lines in service, versus 12,000 for Northpoint and 6,500 for Rhythms. As of the end of 2000, FCC reports showed that the total number of CLEC ADSL lines was only 162,255 (8.2% of the total), whereas CLEC non-ADSL high-speed lines totaled 212,918 (20% of the total); that latter number included T1 and fiber-optic services, as well as DSL technologies other than ADSL. Clearly, the ILECs were keeping the lion's share of the business, even as their competitors were spending hundreds of millions of dollars to compete with them and with each other.

Having a three-way nationwide race for the competitive DSL market would be bad enough, but of course there were many other competitors. Although neither AT&T nor MCI aggressively marketed ADSL during the late 1990s, Sprint took an innovative approach that integrated voice and data. Called integrated optical network (ION), it used ATM technology in conjunction with ADSL to deliver a single-rate nationwide telephone service, as well as Internet access.

A number of regional and local DSL providers also joined in the fray. Nowhere was the competition as intense as in New England, where overexpansion reached tragicomic proportions. While Covad, Rhythms, and Northpoint were already setting up shop, several smaller operations began to compete and received venture capital that allowed them to "get big quick." And although many small, local CLEC DSL providers did, in fact, survive the shakeout and even achieve a degree of prosperity, those that valued expansion over profits met a sadder fate.

Vitts Networks was a small DSL provider based in Portsmouth, N.H., which by 1998 had about a dozen collocations in that state's little seacoast region. Then it received an infusion of venture capital, which brought with it new management and a mandate to expand. It built a new $3.5 million [5] gold-plated data center in a renovated mill in Manchester, grew to 300 employees, and began to expand southward into Massachusetts. Again aiming at business customers, it jumped headfirst into a market that was already getting crowded. Although it reportedly grew to 3,000 customers [6], including the New Hampshire state government, its Massachusetts collocations had far fewer subscribers, on average, than its older New Hampshire ones. It may well have been at break-even on its New Hampshire operations, but the allure of the larger market to its south was, quite literally, overwhelming.

Digital Broadband Communications was based in Waltham, Mass., and had an even more dramatic business plan, building collocations into almost every Bell Atlantic central office in the state, including the smaller ones with only a few thousand lines. It won contracts from the state and from potentially large organizations such as the Massachusetts Association of Insurance Agents

[7], but these were far from adequate; it averaged only a few subscribers per col-location and built several dozen that had acquired literally no customers during the company's short lifetime.

Rounding out the situation was HarvardNet. Originally a small dial-up ISP in the town of Harvard, Mass., about 50 miles from Boston (thus giving it cover against any Harvard University complaints about its name), it received an infusion of venture funding and moved to Boston, where it built the obligatory data center. It entered the Web hosting business, and by June 2000 had installed DSL in 200 Bell Atlantic central offices [8]. It grew to 480 employees serving its 2,000 customers [9] in 85 cities and towns. Its original management team was replaced; the new CEO was an experienced salesman with little operational experience.

So by mid-2000, competition for the DSL market in the Boston area was more than a little intense. Covad, Northpoint, Rhythms, Vitts, Digital Broad-band, and HarvardNet were racing to overbuild each other. All the while, a handful of smaller local CLECs were also picking central offices in which to offer DSL service, sometimes in conjunction with voice. Yet the incumbent, which was by then adopting the name Verizon, had more customers than all of them put together. So when the dominoes began to fall, they fell fast.

Dropping Like Flies

On December 6, 2000, HarvardNet announced that it would be exiting the DSL business. It gave its customers only a few weeks' notice, shutting down on January 15. The Web hosting assets were acquired by Allegiance Telecom, a Texas-based CLEC [10]. On December 27, Digital Broadband Communica-tions filed for Chapter 11 protection, also shutting down its network the follow-ing month. Vitts made a last gasp at survival, hoping that it might benefit from having fewer competitors, but it too filed for Chapter 11 on February 7, 2001; its network was shut down by May.

Northpoint's exit was a bit more complex. The company had agreed, in August 2000, to be acquired by Verizon for $800 million. Although the two were competitors in some areas, Northpoint had substantial assets in areas where SBC was the ILEC; it promised to provide Verizon with a quick footprint in the territory of its supposed archrival. This was the exit strategy that many CLEC investors had dreamed of. "God made CLECs to get acquired," quoted one ven-dor [11] But it was only a brief engagement, never a marriage. Verizon backed out of the deal in December, citing Northpoint's deteriorating finances. Northpoint filed suit, asking for damages if the acquisition did not go ahead, but Verizon did not change its mind again. "The biggest mistake I made was believing Verizon's top management [12]," said Northpoint CEO Fetter after Northpoint filed for Chapter 11 protection on January 16, 2001. Its network was shut down in March; AT&T acquired the bankruptcy assets but did not operate the DSL network. In July 2002, Verizon settled the lawsuit for $175 million, a small fraction of what the aborted acquisition would have cost.

Rhythms did not file its own bankruptcy until August 2, 2001. Its assets were picked up by WorldCom, which at that point was still booming; service continued to many of its customers. Covad filed for Chapter 11 protection two weeks after Rhythms. But as the oldest and most successful of the nationwide DLECs, its fate was different. It emerged from bankruptcy that December and continued to operate, remaining a publicly traded company with a significant market presence.

One might come to the conclusion that DSL was a bad bet, that CLECs should not even attempt to compete with ILECs in this space. But that would not always be a good generalization. As with many other young industries, the earliest investors often lost out, even as the overall business grew. The largest and most visible DLECs went bankrupt, but some smaller ones, with less capital to squander, did not, and even postbankruptcy Covad benefited from the reduced competition. And the bankrupt DLECs left behind a legacy of equipment that they had bought and, in some cases, even paid for without ever opening the box. This went onto the secondary market, where surviving and start-up CLECs were able to purchase it at bargain prices. This made DSL affordable in more places, such as small-market and rural central offices where a family-run Internet service provider might have opened its own CLEC subsidiary. Some of this gear may have been sold, like a lot of other detritus from the boom, on eBay, which is one dotcom that may have actually benefited from the meltdown.

Survivors Face the ILECs' Regulatory Might

And so this could have turned out to be a happy tale, at least for the CLEC sector, after all. But by 2001, a new administration in Washington had appointed a new FCC, and the regulatory environment was changing. Perhaps spurred to some extent by a financial community that felt betrayed by competitive markets, the new FCC, under Michael Powell, took a much less friendly approach to competition. He seemed to view ILEC facilities as the ILEC's ordinary private property, to be used only as they wished, and he even proposed removing the most rudimentary common carriage obligations from their higher-speed services, including raw DSL (a telecom service sold to ISPs; ILEC-affiliated ISPs were, at least nominally, separate arms-length affiliates).

The major area of contention for DLECs concerned shared loops. The Kennard FCC had authorized line sharing in 1998, opening up consumer ADSL to competition. In 2002, however, the D.C. Circuit Court of Appeals vacated this [13], requiring the FCC to revise its rules or offer justification. The court said that the FCC "completely failed to consider the relevance of competition in broadband services coming from cable (and to a lesser extent satellite)." This was a bizarre ruling, since cable does not offer an equivalent service (loops) to CLECs or even offer service to ISPs that buy competitive DSL; cable only competes with the ILECs' *retail* DSL subsidiaries for unregulated *information* service. One might conclude that the CLECs did not argue the case well; on the other hand, it was the FCC, not the CLECs, defending the rule that an earlier FCC had passed. The court's order was stayed until January 2003, at which point the FCC was due to issue its Triennial Review Order (TRO); its UNE rules were subject to periodic review anyway. The FCC could then either provide clearer justification for the rule or remove it.

When the TRO was adopted [14] in February 2003, line sharing was removed; in the final order, adopted six months later, existing shared lines were grandfathered, whereas new shared lines were to be available for only one more year, their price ratcheted upwards over a three-year period. Only one of the five commissioners, Kevin Martin, actually wanted line sharing removed; two others, Jonathan Adelstein and Michael Copps, stated that they were voting with Martin to remove line sharing *in exchange* for his support for UNE Platform, the means by which CLECs resold voice services gotten from ILECs at a cost-base wholesale rate. DLECs were thus collateral damage in a political deal. Some DLECs took the bait and teamed up with voice CLECs, or added voice to their mix, in order to cover the higher cost of unshared loops. But a burgeoning business model had essentially been killed off by the ILECs' regulatory influence. It did little to help the investment climate for competitive providers.

Ethernet LECs Were Data CAPs

Just as the competitive access providers (see Chapter 7) were able to spend money prodigiously by laying fiber in the ground with the hope of selling telecommunications services to large business customers, a new breed of companies sprung up in the late 1990s with the goal of doing the same thing with pure data services. By bypassing SONET and using so-called Ethernet protocols instead, these companies convinced investors that they had created a viable lower-cost business model. They were not CLECs, which were by contrast wedded to old-fashioned telecom models; they were more often called ELECs, or Ethernet LECs. They were going to provide large businesses with the type of bargain Internet service that DLECs provided to homes and small business. Of course it did not quite work out very well for the ELECs, just as it did not work out very well for the CAPs and DLECs.

It may be useful to digress into the magic behind the word "Ethernet." What the ELECs provided was a far cry from the original local area network invented at Xerox PARC in 1973. Ethernet was designed as a way to link computers around a room or building. The first multivendor Ethernet standard, agreed upon by Xerox, Digital Equipment Corporation, and Intel in 1980, ran at 10 Mbps [15]. The name was an allusion to the "ether" of radio broadcasting; Ethernet broadcast bits onto a shared coaxial cable. Individual interfaces then listened to the 48-bit media access control (MAC) addresses on each frame, ignoring the ones that were addressed to others. Transmission depended on an arbitration protocol called Carrier Sense Multiple Access with Collision Detection (CSMA/CD). This was very clever; it allowed any station on the LAN to transmit when it had data, but every station had to simultaneously listen to see if another station transmitted at the same time. If such a collision occurred, each station would back off a *random* length of time before retransmitting. It worked brilliantly well and was amenable to mass production. But it had firm limits. The distance between stations on a given coaxial Ethernet was strictly limited by the speed of light: No two stations could be farther apart than the distance that a bit could travel during the duration of a minimum-size frame (64 octets). Otherwise, collisions would be missed. *Bridges*, packet switches working at line speed at the MAC address level, were needed to relay frames in order to form an extended network.

Ethernet was a technology star of the 1980s. It beat off a challenge from IBM, which promoted the Token Ring LAN as an alternative. It evolved to support unshielded twisted-pair desktop wiring, typically collapsing the CSMA/CD backbone into a small box (hub), which could be bridged to other hubs. It evolved into a fully switched form, wherein the hub became a multiport bridge, no longer sharing a single 10-Mbps pool of bandwidth among its ports. And it fended off almost all competition. In the late 1980s, a 100-Mbps fiber-optic LAN called fiber distributed data interface (FDDI) was designed as an upgrade, but it was slow to market and did not catch on. In the early 1990s, ATM networks, originally proposed by the CCITT for broadband ISDN public networks, had a flurry of interest as a LAN backbone. They too failed: The killer LAN technology of the mid-1990s was Fast Ethernet. This usually was not even based on Ethernet's key CSMA/CD technology, although CSMA/CD Fast Ethernet hubs were possible. More often, it was deployed as a bridged or routed backbone, using Ethernet's frame format on 100-Mbps point-to-point links. A few years later, gigabit Ethernet became available, providing LANs with an upgrade path using the familiar framing format, albeit without a shred of CSMA/CD. By bridging together compatible frames at any speed, Ethernet was able to connect lowly 10-Mbps desktops to fast gigabit servers. And its fundamental simplicity kept the cost below competing technologies.

Ethernet LECs parlayed the "cool" factor of Ethernet with its perceived low cost. Start-ups such as Cogent, Allied Riser, and Yipes took in well over a billion dollars of investor money [16] during the late 1990s boom. Sometimes they were able to acquire fiber from existing CAPs; in other cases they trenched their own. Some companies issued stock, which rose with the boom. This was a tool for expansion in an unusual way: One of the problems the ELECs had was getting landlords' permission to run cable within large office buildings. In order to grease the skids, some ELECs gave landlords stock in exchange for access.

From a customer perspective, the ELECs were a good deal. Ethernet, after all, was supposed to be cheaper than "telecom" circuits. And during the boom, profitability was so far away that the cost of providing service was not a major issue. Cogent was known as a price leader, offering 100-megabit Ethernet connections for $1,000 per month. Of course this was not the same as backbone ISP bandwidth: As with DSL, Cogent's average customer's usage was a low percentage of the peak rate, so its backbone connections could be heavily oversubscribed. Cogent was able, in early 2002, to acquire bankrupt backbone ISP assets of PSInet, reducing its dependence on other providers and giving it a long-haul backbone as well as a more diverse product base.

But the bubble burst hard for the ELECs too. Allied Riser was in deep financial difficulties when Cogent acquired it in late 2001 in exchange for a small amount of stock and the assumption of $123 million of its debt [17]. Yipes filed Chapter 11 in March 2002. Its assets were acquired for a fraction of their original cost, and the new investors continued to operate them under the Yipes name. Cogent did not declare bankruptcy, but it restructured its debt in 2003; its largest creditor, Cisco, ended up with 18% ownership, but more than $200 million of Cogent's debts were forgiven [18]. It continued to operate, even trading publicly, but profits remained elusive. New investment in the ELEC sector has been minimal.

Endnotes

[1]   The original HDSL required two copper pairs (four wires) to achieve DS1 speeds at these distances; newer HDSL2 equipment achieves this on a single pair.

[2]   This is a guideline called carrier serving area (CSA), adopted by the RBOCs; it calls for fiber-optic feeders and remote terminals (digital loop carrier) to serve clusters of subscribers more than 12 kilofeet from the wire center. The FCC uses this as the basis of its "forward-looking" cost formulas.

[3]   A few ISPs had deployed limited amounts of DSL using dedicated copper loops obtained under "alarm circuit" or similar tariffs; once the LECs realized the value of DSL, they often moved to block such activities, and only CLECs could obtain the needed loops. Some isolated pockets of alarm-circuit DSL have reportedly persisted "under the radar" of ILECs.

[4]   Tele.com, Sept. 6, 1999, www.teledotcom.com/article/TEL20000821S0078.

[5]   Haley, C., "Vitts' Bankruptcy Begets Bargain," InternetNews.com, Aug. 1, 2001, boston.internet.com/news/article.php/857741.

[6]   Fosters Online, Jan. 21, 2001 (Dover, N.H., Foster's Daily Democrat).

[7]   "Digital Broadband Gets Pact from MAIA," *Boston Globe*, March 24, 2000..

[8]   "HarvardNet Reaches DSL Milestone With 200th Central Office," PR Newswire, June 27, 2000.

[9]   "HarvardNet to Offer DSL Until Jan. 15," *Boston Globe*, Dec. 8, 2000.

[10]   Allegiance itself went bankrupt in 2003. It was acquired by XO Communications in 2004.

[11]   *Communications Today*, Aug. 9, 2000, reporting on an unattributed quote made by Keith Higgins of DSLAM vendor Copper Mountain, a key Northpoint supplier.

[12]   Taub, S., "Three Companies Default on Debt," CFO.com, January 17, 2001, http://www.cfo.com/article/1,5309,1953,00.html.

[13]   *USTA v. FCC*, 290 F.3d 415 (2002), pacer.cadc.uscourts.gov/common/opinions/200205/00-1012a.txt.

[14]   This "adoption" at an FCC meeting was oddly tentative because the Order that was adopted had not actually been written and was not released for a full six months.

[15]   The IEEE 802 committee was created to standardize LANs. Although an 802.3 standard for Ethernet was created and saw some use, the TCP/IP protocols normally stuck to the earlier frame header, which became the survivor. Later, the 802.3 Ethernet subcommittee went on to standardize many Ethernet-family interfaces, including twisted pair and higher speeds.

[16]   Beck, R., "Ethernet Hype Meets SONET Reality," *Network World*, March 25, 2002, http://www.nwfusion.com/columnists/2002/0325beck.html.

[17]   Bounds, J., "Creditors Ask Chapter 7 for Allied Riser," *Dallas Business Journal*, April 19, 2002,http:// www.bizjournals.com/dallas/stories/2002/04/22/story2.html.

[19]   Fitchard, K., "Cogent Restructures, Raises More Capital," Telephony Online, June 20, 2003, http://www.telephonyonline.com/ar/telecom_cogent_restructures_raises/.

9

CLECs' Winning Strategies Are Met by Rule Changes

Although the Telecom Act had a wide-ranging impact, the single biggest change it effected was, no doubt, the opening of local telephone service to competition. As discussed in Chapter 8, this allowed the digital subscriber line sector to be developed more rapidly by CLECs than it would have been had it been left to the ILECs themselves. And by authorizing competition in almost all aspects of telecommunications, local service competitors were able to try to compete in many other niches. The Telecom Act was, in many respects, a "big bang" for telecom competitors. Many companies, including the large long distance carriers, had been anticipating the Act's passage, following its course through Congress. And the Act itself gave the FCC very short time frames in which to create the regulations necessary to implement its many changes.

But the opening of local competition did not lead to the same rapid dilution of the dominant providers' market share as resulted after the 1984 breakup of the Bell System. Although CLECs did get a quick jump in some specific markets, they had a difficult time navigating the rocky waters of the Telecom Act's ambiguities and the rapid changes in the *effective* laws that resulted from a continuing series of legal battles. DSL providers were not the only ones that found the rules changed just as they were beginning to get back on track. The real winners were the incumbents, which gained a fig leaf of competition without accepting serious market share losses.

The Telecom Act Anticipated CAPs and Resellers

Before the Telecom Act, the closest thing to telecom competition in most states was the federally authorized CAP sector. Competitive access providers had pulled fiber in most top markets and were ready to add local dial tone to their service portfolios. Congress was aware of the CAPs when it passed the Telecom Act; they were certainly assumed to be one of the major models by which local competition would be spurred. Indeed, some states had allowed local competition before the Telecom Act. CAPs were functioning as CLECs in New York, Maryland, and Massachusetts in 1995. New York had granted TCG its first local interconnection rights in 1989. Maryland granted MFS Intellinet's application in 1994, setting an interconnection rate of 6.1 cents per call [1].

Total Service Resale Had Little Value or Margin

The other early avenue for competition was resale. New York's Rochester Telephone actually proposed this in 1993, under its "Open Market Plan." That plan turned the regulated network provider into a regulated wholesale service provider, initially referred to as R-Net, whereas retail accounts were turned over to an unregulated entity (R-Com) that would face competition from other resellers of R-Net's services. The wholesale discount was only 5%. (Both entities, of course, were subsidiaries of holding company Frontier Corp., so if the discount were a bit low, the money would remain in the same stockholders' hands.) Although several companies did set up shop as resellers, they had difficulty making a go of it with such a narrow margin. They also found that the most likely customers to sign up for resale were those who were denied credit by the incumbents [2].

Nonetheless, resale of incumbent local exchange carrier tariffed services was specifically called out in the Telecom Act. The Act essentially defined two types of CLECs—resellers and facilities-based providers. Most states created separate certification programs for the two types of CLECs, with many early applicants only authorized for resale, which appeared to many CLECs, for a time, to be a viable business model. Resellers *only* dealt in ILEC-tariffed services, getting a discount based on the avoided costs of marketing, billing, and collections. Some large-scale reseller CLECs such as USN Corp., which had a costly nationwide rollout shortly after the Telecom Act took effect, failed quite quickly; there was little net profit in essentially acting as commissioned sales agents for the incumbents. All other CLECs were deemed *facilities-based*. That category ranged the gamut, from CAPs and cable companies whose networks were merely interconnected to the ILECs', to UNE Platform providers whose networks were *physically* the same as resale but which purchased their wholesale services at cost-based rates, and a range of options in between those extremes.

State Commissions Had to Administer Federal Rules

The Telecom Act created a most unusual mix of state and federal authority. It set uniform national standards for at least some aspects of competition policy but left the administration of them to the state utility regulatory commissions. This created a highly unusual situation in which federal courts could directly review some state regulatory decisions.

In 1996, in its initial attempt to implement the law quickly, the FCC selected a methodology for UNE prices, called TELRIC. This was based on "forward-looking" costs of incremental capacity, on a hypothetical network built from scratch using the standard methodology of the day, such as a 12,000-foot loop limit supplemented by digital loop carriers fed by optical fiber [3]. TELRIC also took into account the cost of money, in order to ensure that the UNEs netted a fair profit to the ILECs that provided them, and contributed a share of the ILEC's common costs. In other words, it was hardly confiscatory. But in some cases, it produced costs that were well below the *embedded* average cost of the existing ILEC networks. For example, the price of central office switching was falling, but most switches already in the field had been bought at the higher historic price levels. TELRIC only took new costs into account. And in the case of fiber-optic transmission networks, TELRIC did not cover the original cost of laying fiber, just the relatively low incremental cost of lighting up more capacity. Thus the methodology was not popular among ILECs. Nor was it easy to administer, with hundreds of variables that required judgment calls. But then every telephone company cost study methodology since time immemorial (or the invention of the telephone, whichever came last) has necessarily been complex. And it has never been easy to select the right one for any given purpose.

The FCC required states to administer UNE and interconnection prices according to federal TELRIC guidelines, but the initial studies took quite some time to complete. In the interim, the FCC drew up interim *proxy* prices for unbundled network elements. These were based on studies that it had commissioned, such as the *Benchmark Cost Proxy Model*, a massive set of Excel spreadsheets that could predict the cost of loops, switching, and transport anywhere in the country down to the wire center level. But a court struck down the proxy prices because they were not set by the states, so CLECs waited until states had finished their studies before learning the actual prices for UNEs. Virtually all major FCC decisions involved in administering the Act have been challenged in court, with mixed results. This has helped create regulatory uncertainty that in turn has made managing a CLEC far more difficult.

Unbundled Network Elements Reduced Capital Needs

A "facilities-based" CLEC could own all or none of its own equipment, or anything in between, though only certain combinations made sense. And certain unbundled network elements were unquestionably required for widespread competition to exist. At the top of the list was the local loop, the raw copper (or equivalent service derived from a multiplexing arrangement such as a digital loop carrier system) between the subscriber and the central office. Right behind this was dedicated transport, otherwise known as the interoffice facility (IOF). This was the raw bandwidth between central offices, used to access the collocations spread out across a LATA. DS1-speed and DS3-speed local loops, used by business, were also obvious UNEs.

UNE loop rates were generally disaggregated into *density cells*: Loop costs are lowest in high-density urban areas, so UNE loop rates were also lowest there. Urban loop rates set by the states based on their own TELRIC studies were typically in the vicinity of $8 to $12 per month, whereas suburban rates were more often in the $12 to $18 range. Rural rates were higher. Although Bell company rates rarely got much above $25, some non-Bell companies, notably many of Sprint's local operations (the former United Telephone and Centel), set their highest rates over $100 per month! Local telephone tariffs, on the other hand, tended to be most costly in urban areas because they had more lines in their local calling areas and thus a higher "value of service." It did not take a mathematical genius to figure out that competition using UNEs would thus be most profitable in urban areas and unprofitable in many rural areas. (It also begs the question of how rural ILECs can be profitable while charging less than urban ones, but then the industry is rife with subsidies.)

Plain old copper loops were not, of course, the only UNEs. The usual list of UNEs available from RBOCs included many other elements. Besides plain loops, there were ISDN loops, which sometimes had electronics needed to deliver ISDN beyond ordinary loop range, and there were DS1-speed loops, typically deployed using telephone company HDSL hardware, for so-called "T1" services. Dark fiber was also made available both as a loop and as an interoffice facility, albeit, like all UNEs, on an as-available basis. Raw interoffice bandwidth between ILEC offices (dedicated transport interoffice facilities) was also available as a UNE, generally at mileage rates that were a tiny fraction of special access tariff rates. For example, with most special access DS1 rates in the $20 per mile range, UNE DS1 mileage tended to run between a quarter and two bucks a mile. These usually required collocation to access. Operator services, directory assistance, Signaling System 7 ports, the line information database, multiplexors, and various other services were at least for a time available as UNEs, at TELRIC rates.

Section 271 of the Act, which gave the RBOCs an avenue to enter the interLATA business, had its own 14-point checklist, which in some ways was a reiteration of the list of UNEs that was not actually specified elsewhere in the Act. Section 271 required RBOCs to provide CLECs with loops, interoffice facilities, local switches, operator services, and pretty much everything else needed to be a CLEC. It did not independently specify a pricing standard, though. For that matter, neither did the Act; it was merely assumed, at first, that the 14-point checklist would have TELRIC rates applied.

Initial Strategies for Serving "Classical" Voice Business

Some of the first CLECs to spring up after the Act adopted a business model that seemed to fit squarely into the Act's mold. They provided local telephone service in direct competition with the ILECs using their own switches and others' transmission facilities. These transmission facilities could come from the ILEC as UNEs, or from a CAP. Sometimes it was called "Smart Build," because it attempted to optimize the use of capital while building a network. This became known as UNE-Loop operation when it used ILEC loops.

"Smart Build" and UNE-Loop CLECs

In order to access retail and small-business customers, most CLECs that had their own switches still depended on ILEC local loops. This imposed some corollary expenditures. To access these loops, the CLECs needed to establish collocation in the ILEC central offices and obtain dedicated transport facilities from their own switches to these collocation nodes. They then put multiplexing hardware, similar to digital loop carrier hardware, into these collocations. Note that CLECs could *not* install their own switches in ILEC central offices—this was not permitted! They were only allowed to install multiplexors and other equipment needed to make use of unbundled network elements. Even if a CLEC had plenty of extra space in its 100-square-foot collocation cage, it could not put in switching there. CLECs sometimes built their own switch facilities and sometimes put their switches in neutral carrier hotels.

In 1996 and 1997, CLECs usually needed to use collocation cages to access loops. Caged "physical" collocation was costly, though. Such a steep entry price limited collocation to larger central offices where there was some hope of garnering enough business to pay it off. Collocation also carried rental charges, which for a cage was often on the order of $1,000 per month, plus substantial charges for DC power, cross-connect cabling, and other necessities.

By 1998, cageless collocation had become available, and CLECs that had invested heavily in cages during the post-Act big bang were left at a disadvantage to latecomers that could take advantage of cageless. But cost was not the only advantage of cageless collocation; it took up far less space, so the central offices were less likely to run out of space. By the time the boom ended, many urban and suburban COs had indeed run out of room, although by 2001 quite a bit of space had opened up as early entrants failed. The FCC was also of assistance, creating a rule by which CLECs could request a walk-through of ostensibly full buildings, permitting them to identify vacant space that the ILEC may have thought unavailable. The FCC also allowed some limited degree of switching into collocation, provided that it did not require additional floor space. This mainly allowed incidental switching (such as emergency stand-alone mode) within a remote line terminal, and the data switching devices usually called routers, which were often needed to join multiple shelves of equipment, such as DSLAMs, to interoffice transport.

Many smaller ILEC buildings, especially in outlying areas, had virtually no room for collocators to begin with. Suburban and urban ones usually started out with cavernous vacant space, often used for offices or, reportedly, table tennis. This was because central office switches of the electromechanical era, such as crossbars and steppers, were far larger than the digital switches that took their place. So if the large COs were full, then one might expect that competition was robust and competitors were selling well. But it was not that way at all. Most collocation cages were nearly empty, with one or two equipment racks and maybe some spare parts. Most collocation racks were under-utilized: There was usually far more capacity present than there were customers. Many CLECs during the boom years invested heavily, using cheap capital, expecting a far larger market share than they ever attained. The voice CLEC business in this regard was not much different from the DLEC sector. And of course some CLECs had both voice and DSL services.

UNE Platform Displaced Resale and Discouraged UNE-Loop

The FCC, during the boom years under Chairman William Kennard, had a liberal interpretation of some of the Telecom Act's more mystifying words. For example, Section 251(d) of the Act said:

> (2) ACCESS STANDARDS—In determining what network elements should be made available for purposes of subsection (c)(3), the Commission shall consider, at a minimum, whether:

> (A) access to such network elements as are proprietary in nature is *necessary*, and

> (B) the failure to provide access to such network elements would *impair* the ability of the telecommunications carrier seeking access to provide the services that it seeks to offer.

> [Emphasis added.]

This became known as the "necessary and impair" test. The Kennard FCC held that almost *all* elements used in most telephone service, though not packet-switched data equipment such as DSLAMs or auxiliary nontelecom features such as voice mail, should be unbundled at TELRIC rates. Thus the entire analog telephone line, including switching, could be ordered as a set of UNEs. This combination became known as the UNE Platform (UNE-P). ILECs, not surprisingly, hated it and fought it everywhere they could.

The Act said that elements could be combined; ILECs argued that it was the CLECs that had to do the combining, in their collocation cages. This was particularly illogical with regard to local switching and its associated elements such as shared interoffice transport. The latter element, by which CLECs using UNE-P paid for the usage of the trunks interconnecting the switches on the ILEC's local network, did not exist as a physical entity. It was *shared*, after all, and hard-wired to the ILEC switch. Nonetheless the Eighth Circuit Court of Appeals initially supported the ILECs view. UNE-P thus did not really have a chance to take off until after the Supreme Court's 1998 ruling in the *AT&T v. Iowa Utilities Board* case overturned the Eighth Circuit, affirming the viability of UNE-P as well as of the FCC's TELRIC price formula. It was a landmark victory for CLECs.

UNE Platform, with its wholesale price to the CLECs not coupled to ILEC retail tariffs, was cheaper than total service resale in urban areas, especially for business subscribers. It was usually even cheaper for residential service, at least once a few options were figured in. That was another difference between cost-based rates and resale; because features like call waiting and caller ID did not cost much to provision, they did not cost much as part of UNE switching. The ILECs' rate structure was thus in serious risk of being shattered. The ILECs did not embrace this excuse to migrate to a more cost-based rate structure themselves; rather, they continued to apply political and legal pressure to shut down UNE-P. After receiving Section 271 authority to offer long distance, RBOCs took competition a bit more seriously, introducing more competitive bundles of local and long distance service. Of course it was the big long distance carriers, AT&T and MCI, that most strongly embraced UNE-P, largely because of one other detail that distinguished it from resale. Under UNE-P, the CLEC was the carrier of record, not the ILEC; the CLEC was merely outsourcing a number of tasks (well, an awful lot of them) to the ILEC. And thus the long distance carriers owed their switched access charges to the UNE-P CLEC, not the underlying ILEC. When the long distance carrier *was* the CLEC, it was thereby able to avoid a substantial part of its own expenses by paying them to itself.

Some of the UNE-P CLECs were in the Centrex business, selling complex business telephone systems using ILEC switches. This was one of the first items to be challenged under the necessary and impair test. Centrex competed directly with PBX systems, which were widely available, so it was evident that competition for that market could exist without ILEC switching. Primary rate ISDN and channelized T1 services, used for both PBX trunks and modem pools, were also relatively competitive because, by 2001, there were many central office switch vendors capable of supporting those services, and prices for T1 switching ports had fallen to a fraction of their pre-1996 levels. CLECs had spurred switching technology; the ILECs could argue that in most major markets, there was plenty of CLEC switching capacity available. CLECs could, after all, buy from each other. It was only a slight problem that most CLECs did not choose to sell wholesale to each other or did not bother to create mechanisms for the purpose. CLECs looked at each other as competitors when they could have focused on the ILECs, whose market share was still many times their own. Not that wholesaling was always easy: CLECs typically used loop carriers based on the GR-303 standard, which were built to talk to a single switch and thus could not often directly connect to two different CLECs' switches.

EELs Created an Opportunity to Serve Businesses

In 1999, the FCC created a rule by which unbundled dedicated interoffice transport—cheap T1 pipes—could be ordered in combination with an unbundled DS1 loop, to any of the ILEC's central offices in the LATA, without requiring collocation at the subscriber's serving CO. This allowed, for example, a CLEC in Newark, N.J., to provide T1-based service to customers in Hackensack, Hoboken, and Hohokus without requiring collocations in those localities. This combination of transmission UNEs was called an expanded extended loop, or EEL [4]. ILECs that refused to provide EELs had to provide unbundled DS1 switching in major markets instead; EELs became almost universally available. But EELs came with a catch. The lines, in order to qualify, had to carry a "substantial" amount of "local" voice service, rather strictly defined and requiring written certification of compliance by the CLEC. This was to prevent special access applications such as ISP access and WATS-style long distance lines from becoming EELs. CLECs could, however, provide combined voice and data services across them; the savings compared with special access sometimes made the voice service virtually "free" to the end user. A number of CLECs pursued the EEL avenue in the 2000s to considerable success.

The ISP Dial-In Business and CLECs: A Match Made in Heaven

The market sector in which CLECs were most successful was, without a doubt, the provision of incoming dial-in capacity to Internet service providers. The Telecom Act was passed during the steepest part of the Internet boom. Consumers were buying modems by the millions. ISPs were hard-pressed to keep up with demand. And ILECs were particularly hard-pressed to keep up with demand for the ISDN PRI lines that worked best with remote access servers, the systems that combined high-capacity digital line interfaces, modems, and routers. For some years, ILECs had, after all, been growing only modestly faster than the population. They had been accustomed to upgrading their network capacity based on lengthy (up to five-year) forecasting intervals, and their PRI rollouts had been leisurely at best. Suddenly, they were being asked for PRIs by the dozen. And they were not always in the most populous places; ISPs wanted dial-in number with the largest number of possible users in their local calling area. That did not always line up with where the ILEC had the most capacity; it might be a suburban area local to a large urban area as well as other suburbs.

Some of the ILECs reacted to the ISP boom by trying to stamp it out. Bell Atlantic submitted an ex parte white paper to the FCC calling for the imposition of switched access charges on ISP-bound calls (the so-called modem tax). Its paper cited the cost of adding hundreds or thousands of lines of capacity to some of its oldest analog central offices, whose local calling areas happened to be particularly desirable. Bell Atlantic made no attempt to engineer a more economical solution: If the subscriber wanted numbers in a town where it had a 1AESS electromechanical switch, then by gum it would use that switch—even if, as was usually the case, the lines were *foreign exchange*, actually delivered to the ISP elsewhere. Pacific Bell also decried how Internet growth would cause it to have to spend $100 million more on capital upgrades in 1997 than it had planned. That sounds like a lot of money to the average consumer, though it was only a modest percentage increase given the size of Pacific Bell's California market and the amount of growth in the state's economy that year.

But the FCC did not stamp out the Internet revolution, and the network did not collapse. Instead, the CLECs rode in to the rescue. CLECs installed dozens of large central office switches between 1996 and 1998 whose primary mission was serving dial-in modems. Some CLEC start-ups were primarily aimed at that marketplace. PacWest in California and Global NAPs in Massachusetts were among the larger examples; a number of ISPs also started their own CLEC affiliates in order to "offload" their modems from the ILEC network.

CLECs had many natural advantages for this market. They could move much faster. They also looked at ISPs as *customers* to be delighted, rather than, as ILECs seemed to, as *ratepayers* that they were being forced to serve. And they offered the ISPs collocation, something ILECs never did. By putting their modems in the collocation rooms at CLEC sites, ISPs' local loops were merely interior cables. This alone often saved hundreds of dollars per T1. Some CLECs opened carrier hotels for ISPs, centered around their switches. These tended to have staying power because they offered a service that independent carrier hotels often lacked and dial-in ISPs needed. Other times, CLECs located their switches in other companies' carrier hotels, so that they could take advantage of available transmission facilities and be less dependent on the ILEC, while still being near a home for their customers' modems.

And when an ISP created a CLEC, its relationship with the ILEC changed. It was no longer a subscriber; it was a peer. The circuits between the two carriers' networks were no longer PRIs, costing several hundred to a couple of thousand dollars per month apiece. They became intermachine trunks (IMTs), circuits interconnecting the two carriers that each carried half of each call. And under the FCC's rules for CLEC interconnection, IMTs were paid for by the carrier that originated the call. For ISP-serving CLECs, that was almost always the ILEC. For a dial-in ISP, sticking with the ILEC became a serious cost disadvantage, compared with their competitors that either teamed with, or became, CLECs.

"Virtual NXX" Made Dial-In Available in More Areas

The CLEC sector really did show innovation. One important if controversial one happened almost by accident. CLECs were not allowed to put switches in ILEC central offices, so they installed regional switches that served many exchange areas. This made it possible to offer foreign exchange service across the switch's serving area at no incremental cost. A Santa Monica ISP needing a downtown Los Angeles number from Pacific Bell would get a foreign exchange line, paying for mileage between its site and Bell's switch *in* the central Los Angeles exchange. (And since Santa Monica was GTE's territory, coordinating the line's installation was that much more difficult.) But a Santa Monica ISP getting numbers from a CLEC could get Santa Monica numbers, Los Angeles numbers, San Fernando Valley numbers, and quite possibly the "full house" of greater Los Angeles' many local calling areas, all on the same lines. This aggregation of foreign exchange numbers on one switch for the purpose of creating additional local calling areas came to be known as virtual NXX [5]. CLEC-serving ISPs adopted it in droves.

That, in fact, was a major reason why there were so many area code splits between 1996 and 2000. ISPs and CLECs that served them wanted "full house" coverage across major metropolitan areas. That often required numbers in dozens of rate centers. So a small rate center that might have had a single ILEC prefix code now might have five or six CLEC prefix codes, most of them using only a handful out of their 10,000 numbers. This could have been alleviated in several ways. ILECs could have consolidated rate centers, giving larger local calling areas. But that would have cost them short-haul toll revenues. States could have allowed CLECs to have "oddball" prefix codes that were local to the entire LATA or metropolitan area, but that would have given them an apparent advantage over ILECs. The ILECs actually did make such prefix codes available under their own tariff, but they were charged at switched access rates: The subscriber paid by the minute to receive calls. That was a nonstarter in the ISP market. So the CLECs got the same result by grabbing large numbers of prefix codes. This led to repeated area code exhaust.

The problem was finally solved when local number portability (LNP) became widely available. The Telecom Act required the industry to develop methods of making local telephone numbers portable, so a subscriber could change carriers and keep its number. This was not, however, all that easily accomplished; competition began before LNP was in place. By the late 1990s, the industry had adopted a standard method for LNP; it then had to be phased in to the network, starting with the largest markets. But even LNP left the CLECs needing "NXX" prefix codes of their own because the ILECs did not willingly issue *new* numbers from its own pools to CLEC subscribers. The real answer finally turned out to be *thousands-block pooling*. This made use of the LNP mechanism to divide a prefix code among multiple carriers. Carriers would then be rationed numbers 1,000 at a time, allowing up to 10 carriers to share a single prefix from which each could assign new numbers to its own subscribers.

Different states looked at virtual NXX differently. It took off and was widely deployed in most of the urban coastal states before anyone really could stop it. But in 2000, the Maine Public Utilities Commission (PUC) ruled that WorldCom subsidiary Brooks Fiber had violated state rules with its "remote exchange" VNXX service. The PUC was upset that Brooks had taken over 50 prefix codes from the state's one area code, while it was only physically located in Portland, refused to serve customers physically outside of Portland, and was in fact only authorized to serve the Portland area. It held that calls to those non-local numbers were, in fact, interexchange and subject to switched access charges. Brooks had to phase out the service and return most of its prefix codes to the North American Numbering Plan Administration. Shortly afterwards, the South Carolina PUC made a similar ruling against Adelphia Business Solutions. Those states were the exceptions, but some other rural states had no virtual NXX CLECs, even absent a formal ruling.

The New Hampshire PUC issued a more nuanced ruling in 2003. It recognized that ISP-bound calls were, according to the FCC, exempt from switched access charges. (Whether the FCC's long-standing "ESP exemption" applied to geographically nonlocal calls was an interesting question that states did not always agree upon.) But it did not want to waste prefix codes for virtual NXX. So it ruled that a CLEC could not have a prefix code rated to an exchange in which none of its subscribers were physically located—the traditional meaning of virtual NXX. But it allowed the CLECs to have a single oddball NXX code for ISP use that would be local to the entire state, the "information access NXX." Had such a rule been adopted in 1996, the country would no doubt have had far fewer area code splits, and the CLECs would still have been able to service ISPs during the dial-up boom. By 2003, though, the number of dial-up users was already in decline.

Besides the prefix aggregation of virtual NXX, some CLECs had another way to service ISPs more economically. They developed *switchless* networks, in which their remote access servers themselves acted as both the switch and the modem. This was pioneered by a Massachusetts start-up, XCOM Technologies, which by 1998 had built a Signaling System 7 gateway between its Ascend TNT servers and the Bell Atlantic network. XCOM had started out using a costly Northern Telecom DMS-500 switch but was able to phase it out [6]. Level 3 bought the still-young company in 1998 for $154 million, an amazing payday for a tiny start-up, but Level 3 did expand the switchless service across the United States and Europe. It did not, however, have to depend on a home-grown gateway. Ascend, which was acquired that year by Lucent, developed its own Ascend Signaling Gateway; Nortel also introduced a gateway for its CVX-1800 RAS.

Reciprocal Compensation Led to Large Initial Profits

The relationship between CLECs and ILECs was very different from the relationship between IXCs and LECs. An IXC was a customer of the LEC, purchasing access services, usually at highly inflated rates, at both ends of the call. A CLEC was, however, a peer, a co-carrier sharing in the delivery of one network's call made to the other's subscriber. There was ample precedent for this in areas where more than one LEC divided the turf. Los Angeles, for instance, was primarily Pacific Bell's franchise area, but it was also home to a patchwork of GTE franchise areas. Those two carriers were tightly interconnected and happily handed off calls to each other. On hundreds if not thousands of routes across the country, Bell and independent LECs, or two independents, handed off local calls to each other. In the vast majority of cases, the arrangement was *bill and keep*, in which neither side paid the other to terminate its calls. (This was a far simpler arrangement than applied to the shared intraLATA termination of toll calls, for which an elaborate procedure called *meet-point billing* applied, and the two carriers divvied up the access revenues based on a formula that took into account which carrier actually owned what share of the interoffice facilities.)

When the Telecom Act passed, most of the CAPs that became the first CLECs wanted to be treated the same as independents, with bill and keep. This made sense to them because their early customers, mostly large businesses, typically made more local calls than they received. High-volume call centers, after all, typically used 800 numbers, which were not *local* interconnection. The ILECs recognized this, and they had experience with cellular interconnection, in which more than twice as many calls were typically originated than terminated by the cellular network. So they fought hard against bill and keep, with emotional pleas like this from Bell Atlantic:

> The most blatant example of a plea for a government handout comes from those parties who urge the Commission to adopt a reciprocal compensation

price of zero, which they euphemistically refer to as 'bill and keep.' A more appropriate name, however, would be 'bilk and keep,' since it will bilk the LECs' customer out of their money in order to subsidize entry by the likes of AT&T, MCI, and TCG.[7]

The FCC's *Local Competition Order* thus imposed a system called *reciprocal compensation*. It hinged on this wonderfully ambiguous wording in Section 251(b)5 of the Telecom Act, describing the duties of "all local exchange carriers" (and thus including both ILECs and CLECs):

(5) Reciprocal Compensation.

The duty to establish reciprocal compensation arrangements for the transport and termination of telecommunications.

The FCC decided that this did not apply to toll calls, because the switched access system was already in place, but that CLECs and ILECs should, in general, pay each other to terminate local calls. The rate that each paid would, however, almost always be the same, so if traffic were in balance, the fees would net to zero.

The aforementioned Bell Atlantic filing suggested that if the CLECs did not like paying reciprocal compensation, or thought that the rate was too high, then maybe they should seek out customers who received more calls than they made, such as Internet service providers. And that is exactly what the CLECs did. Reciprocal compensation helped fund the initial buildout of ISP-serving CLEC network capacity. ILECs had dictated initial reciprocal compensation rates, albeit with state consent; absent cost studies, some of these rates were quite high. Some of the early Bell Atlantic interconnection agreements set the reciprocal compensation rate at 0.8 cents per minute. A single PRI serving ISP modems could easily carry 250,000 minutes per month; at that price, it would generate $2,000 per month of reciprocal compensation. Combined with virtual NXX to ensure a wide local calling area, many CLECs were, by 1998, doing quite well.

ILECs on the Attack

Naturally, the ILECs were not happy about this situation, the monster they had unwittingly created. And they were certainly not going to actually *compete* for the ISPs' business, not when they had been calling for all such calls to be switched access. But there were a couple of exceptions. One ILEC, Connecticut's independent Southern New England Telephone, not yet owned by SBC, built a statewide "full house" of its own for ISPs, using a dedicated offload switch. GTE, not yet owned by Bell Atlantic/Verizon, also had an offload service, CyberPOP; while covering its own major territories, it was largely targeted at the many Bell company subscribers who were within local calling range of a GTE switch. Bell Atlantic's own offload service, Internet Protocol Routing Service (IPRS), seemed to be primarily designed for the use of its own ISP subsidiary. No, the ILECs were not going to take the CLEC's reciprocal compensation windfall lying down; they were going to fight them the way they knew best, with lawyers.

So by 1998, Bell Atlantic, and soon the other ILECs, were changing their tune. Calls to ISPs were not, after all, local and should not, they insisted, be subject to reciprocal compensation. In some states the ILECs practiced "self-help" and simply stopped paying CLECs that served ISPs. Thus began a long round of state-by-state hearings on the issue. More often than not, the CLECs won at the state level, at least initially, retaining reciprocal compensation in California, Florida, and New York. In February 1999, the FCC held that ISP-bound calls were *jurisdictionally* interstate because the end-to-end payload of an Internet connection was often out of state. This weakened the "two-call" theory, that a call to an ISP was an example of local *telecommunications* whose payload was a distinguishable example of interstate *communications* [7]. But that did not, by itself, end reciprocal compensation. The ILECs then won a major victory in Massachusetts, whose Department of Telecommunications and Energy reversed its earlier position and decided, on the basis of that FCC ruling, that ISP-bound calls should not receive reciprocal compensation [9]. That decision established a rebuttable presumption that if a CLEC's traffic were more than 2:1 inbound, then the excess was ISP-bound. That hand-waving got around the issue of determining just what was an ISP-bound call, since ISPs are not regulated or licensed as such and thus do not technically have to identify themselves to LECs.

The Grandfather Clause Locked Out New Entrants and Squeezed the Old

In April 2001, the FCC issued a new ISP ruling that once again affirmed the interstate nature of ISP-bound calls. It adopted the Massachusetts model of using a traffic ratio, 3:1 by default, to identify ISP-bound calls. But it allowed states to leave in place existing reciprocal compensation arrangements, if they so chose, subject to a cap on the number of compensable minutes. That grandfather clause kept newcomers out of the business; new CLECs got no reciprocal compensation for ISP-bound calls because their cap was zero. Reciprocal compensation rates for eligible CLECs were also capped at a level declining to $0.0007 by 2004. Later that year, the eligible minutes cap was lifted.

And thus by several means, in several steps, the CLECs' single most profitable business was shut down because the ILECs had the rules changed when they realized who was winning. This happened just as the telecom sector in general was collapsing, when the CLECs were hard-pressed to find funding with which to reorient their businesses. It was a portent of further regulatory difficulties to face the CLECs.

New Generation Switching Equipment Lowered Capital Costs

The introduction of local telephone competition after the Telecom Act of 1996 led to the birth of the competitive local exchange carrier sector. Although some of the CLECs were simply resellers, essentially commissioned agents of the ILECs whose services they resold, many facilities-based carriers built their own networks. This created additional demand for telephone equipment.

When the Telecom Act let loose a flood of CLECs, those that planned to offer ordinary telephone service had little choice as to what switch vendors to use. Between 1980 and 1995, the number of central office switch manufacturers serving the domestic market had declined. This had a few likely causes. Computer-controlled switching systems were less costly to build, but it cost as much to develop software for a 3% market share as for a 50% share, and software was most of the cost of a switch. In that era of minicomputers and slow microprocessors, switch software was usually a complex mass of embedded purpose-built code.

So when CLECs began to look for switches, Lucent offered the 5ESS family that it had been selling to ILECs for the preceding 15 years. Its best deal was the 5ESS VCDX, the "very compact digital exchange," which consisted of one node of its modular 1980-era flagship switch, modernized by replacing its proprietary main processor with a Sun Microsystems server. "Very compact" was relative: A small VCDX still occupied multiple bays. Nortel had the well-established DMS-100, which it repackaged for CLECs (same hardware, different bundling of software) as the DMS-500. Both of these had entry-level prices over a million dollars. So did the obvious third choice, the Siemens EWSD, which had also been homologated into the RBOCs' networks. There were not many other choices, because the telephone switching business had become more concentrated over the preceding decade. DSC Communications, founded as Digital Switch Corp. and purchased by French giant Alcatel in 1998, had been selling its DEX tandem switches to long distance carriers; these had limited end-office capabilities but were used by some CLECs.

Venture capital abhors a vacuum, and the coincidence of the Telecom Act and the boom led to a burst of activity. A number of companies jumped in to address various niche markets that had sprung up, largely among CLECs. These new switches were not, for the most part, clones of old Lucent designs; they reflected both the far greater capacities of modern hardware, especially fast semiconductors, and newer software architectures.

Enter the Softswitch

A new word entered the vocabulary during the boom, following the Telecom Act, that described the predominant new idea in telephone switching. In addition to the sudden appearance of CLECs on the scene, the development of Voice over Internet Protocol (VoIP or Voice over IP) was certainly an impetus for some new development, especially the so-called "softswitch" concept. But that term was never well defined. The "softswitch" began as part of Voice over IP architecture: With an IP network, possibly even the whole Internet, functioning as the core *matrix* of the switch, call control itself could be handled by a processor somewhere on the network. This became known as a *call agent*, or softswitch. Initially it was stand-alone software that managed distributed *media gateways* that put voice onto a network—usually IP, but sometimes ATM or even TDM. But then it also came to refer to switches whose internal software had hooks and handles for customization, or even switches whose software was only partially complete, by design, so that original equipment manufacturers (OEMs) or carriers themselves could develop customized solutions. These were both in marked contrast to the monolithic model of older switches, whose extensibility, other than vendor-supplied features, was usually limited to the Intelligent Network capabilities defined by Bellcore. These two categories of softswitch were not mutually exclusive; some of the new switches were not only extensible but had internal architectures that visibly included both a call agent capability (a software application) and a media gateway (the switch backplane and its interface cards).

Most new switches did, in fact, consist of a call agent and a media gateway, but then so did a 1930's vintage crossbar switch. Eventually, even legacy switch vendors took to calling their products softswitches, but the term mostly stuck to the new generation of switches that was to arrive on the market in the late 1990s.

The CLEC Boom and Bust Led to Glut of Vendors

Quite a few central office switch vendors—almost all of who could call their products softswitches—started up or entered the market during the boom. One of the first CLEC-oriented switch vendors began as Digital Telecommunications Inc., before changing its name to DTI Networks. It could trace its heritage back to a mid-1970s PBX start-up, Wescom Switching, but by the 1990s it was a small manufacturer of specialty call distributors based in Tennessee. After the Telecom Act, it rolled out a primitive but inexpensive "tandem" switch, the DXC, which was suitable for PRI offload. Dozens of ISP-CLECs bought them. By 2000, it had moved its headquarters to Florida and merged with DSL equipment vendor Coppercom and taken that company's name. This was a timely bailout for Coppercom shareholders because its core market, SDSL, was horribly oversaturated and almost completely wiped out by the DLEC melt-down. The renamed company released a more modern small switch, the CSX, and stayed independent until 2002, when its cash ran low and it was acquired by a private equity firm, Heico Corporation. It continued to sell to CLECs but also added rural ILECs to its target market.

Sonus Networks was another switch vendor start-up, founded in Massachusetts in 1998 and aimed at the large-scale market. Its core products were primarily marketed as VoIP media gateways, but they also had ATM and TDM capabilities and could comfortably handle large-scale PRI offload applications. End-office applications, however, required third-party software, typically from Broadsoft. Its products met considerable success, taking a leading role in carrier IP networks, though not very much with CLECs; most of its key customers were in the long distance business. Sonus went public early, taking advantage of the boom to go public in 2000. Its stock went near $80 that year but fell by 2002 to, at one low point, about a quarter a share. It eked its way back up, peaking near $8 in early 2004 before slipping as it discovered that previous years' financial results needed to be restated.

Just up the road from Sonus, another Massachusetts start-up, Convergent Networks, also built a large-scale switch, initially with TDM and ATM capabilities and later VoIP. Aimed at PRI offload applications, Convergent—ironically headquartered in the former Wang Tower, a monument to the 1980s meltdown of the minicomputer business—was somewhat less successful than Sonus. It had one large customer, Global NAPs, a CLEC that provided large-scale virtual NXX offload in numerous states. When the company's financial situation became dire in 2002, it was acquired by an affiliate of Global NAPs. This made the Quincy, Mass.-based CLEC an interesting anomaly—a local exchange carrier that owned a switch manufacturer. But it was no Bell System; it was a meltdown bargain.

And down the road in Marlborough, Mass., 1998 start-up Telica Inc. built a large-scale switch with both TDM and VoIP capabilities. Funded, like many of its neighbors, by top-tier venture firms, it had some success in penetrating both ILEC and CLEC markets. It was acquired by Lucent Technologies in 2004 for $295 million, a relatively happy outcome for investors in these companies.

That might sound like a good selection of switch vendors, but they were not alone; they were just the survivors. Siemens, for instance, funded a boom-era start-up, Unisphere, that was going to produce a next-generation switch. It pulled the plug a few years later. Ironically, Unisphere and Sonus once shared a building in Westford, Mass., that had earlier been occupied by minicomputer giant Digital Equipment Corp. A New Jersey start-up, Tachion Networks, planned an ambitious product that was at once a central office switch, a SONET multiplexor, and an IP, ATM, and Frame Relay switch. It heroically planned to build this using 18-layer printed circuits. It did not work. The company folded in 2001. Gluon Networks was building the CLX "converged local exchange," integrating a switch, a media gateway, and a DSLAM in a Taqua-sized package. The Petaluma, Calif.-based company ran out of money; its assets were acquired by bargain-hunter Zhone Technologies in 2004 for a mere $7 million. And one of the industry's earliest post-1996 switch failures was Salix, which was building a large-scale TDM/VoIP switch when it was acquired by equipment giant Tellabs for about $300 million in early 2000. It never came to market; Tellabs pulled the plug in mid-2001, as the Illinois-based parent company was bracing for the meltdown.

UNE-P Dominated CLEC Statistical Growth

Once the meltdown began, investment in new facilities-based CLECs slowed to a crawl. True, the existing capacity of boom-era CLEC switches was very lightly utilized. But absent a developed wholesale business of sales between CLECs, and given the cost of tying CLEC switches to ILEC loops, that capacity was hard to tap into. Competition did, however, at least nominally expand, primarily via UNE Platform. By 2000, CLECs had software that could tap into ILEC operational support systems and place UNE-P orders. In the majority of cases, this switching of carriers was only a change in records. The customer's phone number was unchanged, and the same loop remained on the same switch port. Software vendors even built applications that could take a subscriber's number and convert it to UNE-P with the click of a mouse. A large number of CLECs thus sprung up whose business model was to be a pure play in UNE-P, with marketing and billing but no network of their own. And the big IXCs adopted this model in spades: MCI's pioneering nationwide-local-calling plan, The Neighborhood, was built out of UNE-P. Neither MCI nor AT&T—both of whom had extensive local switching networks for business customers—provided residential service on their own local networks in the early 2000s; UNE-P's margins were higher.

In December 1999, CLECs nominally controlled about 8 million lines, 4.3% of the country's local telephone lines [10]. Of these, about half were provisioned via resale, and under a third (about 2.7 million lines) were on CLEC-owned transmission facilities. That latter category included ISP dial-in circuits delivered locally in CLEC collocations. About 6% of CLEC lines—around half a million—were on the newly minted UNE Platform. About a million lines used UNE-loop; that included the DLECs as well as classical-model voice providers.

One year later, after the last full year of the boom, CLEC-owned facilities had almost doubled, to 5.2 million lines. UNE-Loop had more than doubled, to about 2.4 million lines, but that paled compared with UNE-P, which more than quintupled in size to over 2.8 million lines. Resale lines were being converted to UNE-P in droves; by mid-2001, UNE-P had more lines than resale.

By December 2003, CLECs had a nominal 16.3% market share. But of these, only 16% (4.7 million lines) were resold and 23% (6.9 million) were on CLEC-owned facilities. Cable represented about 3 million lines—the number of lines other CLEC-owned transmission facilities had had barely grown in three years of tight capital. UNE loop had leveled off at about 4 million lines, but UNE Platform, with over 15 million lines, represented more than half of the CLEC line count.

An ILEC-Friendly FCC Throws New Obstructions at Surviving CLECs

But by late 2003, UNE Platform was a shaky foundation indeed to build a business on. The ILEC had made a full-court press against it, and its foundations were starting to crumble. ILECs had never really stopped litigating against UNEs. Although the Supreme Court's 1998 ruling in *Iowa Utilities Board* was a fairly ringing endorsement of the FCC's original position, the ILECs, through their trade organization, the United States Telecom Association (USTA), continued to seek judicial support. In 2002, the U.S. Court of Appeals for the District of Columbia Circuit ruled, in *USTA v. FCC* ("USTA I"), [11] that the FCC had to re-examine its UNE policies giving more strict scrutiny to the "necessary and impair" test. In particular, the FCC needed to take into account the availability of alternative sources of supply other than the ILECs.

This was consistent with FCC Chairman Michael Powell's own view of competition. An outspoken foe of UNEs in general, Powell preached the gospel of "intermodal competition." This meant that competition should be between *the* wireline provider, the cable provider, wireless providers, satellite providers, and, perhaps, power-line carriers. With multiple modes of access to the home, the consumer would have some degree of competition to protect against the most egregious abuses of monopoly. Powell's view had some support from manufacturers because UNEs were always *existing* facilities, used in lieu of new capital investment, whereas intermodal competition theoretically encouraged overbuilding, and therefore encouraged incumbents to continue to invest in facilities that they could *totally* control. In one docket (FCC 02-33), the FCC suggested withdrawing common carrier obligations from ILECs when they provided "broadband" service, such as DSL or T1 data lines. This would permit them to cut off access to all the independent ISPs that depended on ILEC DSL services. Combined with the failure of the DLECs, ISPs would be drowned, leaving ILEC- and cable-captive ISPs with a duopoly. Powell's view was at best only remotely tied to the obvious intent of the Telecom Act, but it had support from some quarters who blamed the meltdown on the Telecom Act and upon the competitors that it spawned. And it made him very popular among the ILECs, which of course also owned the largest wireless carriers.

The Triennial Review Order Fiasco

On February 20, 2003, the FCC announced that it had reached consensus on UNEs. This decision was known as the Triennial Review Order, as the FCC had said that it had planned to review its three-year-old UNE rules anyway. The actual text of the TRO did not come out until six months later! Clearly, the FCC had not reached agreement; the text, more than 500 pages long, was filled with personal jabs between the commissioners. The TRO was a complex mass of compromises. Its key feature was that in many key areas, it passed the issue of determining "impairment," a prerequisite to unbundling, to the states. The final determination on preserving UNE Platform was left to the states, which could make a finding of no impairment. The TRO also passed the key issue of dedicated interoffice transport to the states. It provided guidelines for determining when there was *sufficient* competition, on a route-by-route basis, to permit an ILEC to limit its availability at DS1 and DS3 rates. Similar tests were created for dark- fiber and high-capacity (DS1 and above) local loops. CLECs around the country were then asked to provide their state commissions with details of their network facilities, so that the number of suppliers on each route could be determined.

The TRO also gave the ILECs the vast majority of what they wanted with regard to FTTH. The ILECs' view was that such facilities should not be unbundled at all; they were, in effect, holding long-promised network upgrades hostage for a change in regulation. The FCC went along with that, citing the tortured logic that since the ILECs had to date only pulled a few thousand such lines nationwide, even fewer than had been pulled by CLECs, the CLECs' larger share of that miniscule market demonstrated that no impairment existed. The one concession to CLECs concerned FTTH overbuilds by the ILECs. The ILECs had to continue to make old copper available or provide CLECs with a single voice-grade narrowband path across their fiber.

The TRO was a good deal for the ILECs, but they did not think it good enough. Both the CLECs and ILECs filed suit, in at least three appellate districts. The cases were consolidated at the D.C. Circuit. In March 2004, that court issued its *USTA II* ruling [12]. It vacated the TRO with a scathing rebuke of the FCC's delegation of its impairment findings to the states. It also expressed doubt that *any* UNE Platform was necessary, based on a frankly incomplete reading of the record [13]. And while it acknowledged on the one hand that the Supreme Court had upheld TELRIC rates as legal, it dismissed TELRIC pricing with the amazing tag line, "In competitive markets, an ILEC can't be used as a piñata." It then remanded and vacated the TRO, subject for a brief stay to allow the FCC to *promptly* rewrite it. The Powell FCC did no such thing, of course; its record on remands was notoriously poor, and this was no exception. Instead, Powell called on CLECs and ILECs to engage in negotiation to create "commercial agreements" to replace UNEs. Of course that was farcical—the ILECs held *all* the cards, so, reportedly [14], they mostly offered only access tariff pricing for interoffice facilities, plus UNE-P at substantially increased prices. Very few deals were struck, although Qwest, mindful of competition from cable modems, did strike a line-sharing deal with Covad—with whom it had such a deal *before* it was mandated nationally by the FCC.

Several CLECs did appeal to the Supreme Court but did not get a stay pending consideration. Thus the *vacatur* technically took effect on June 15, 2004. The vacatur did not, as one might expect, mean that the TRO was voided. Rather, because the TRO was itself a remand, it meant that *any* UNE that was in any way contested—practically everything except analog loops— would no longer be required. For ILECs, this was like a month of Christmases! This led to a game of chicken: The ILECs typically invoked the change-of-law clauses in their interconnection agreements and gave 90-day notices to CLECs. Then, in midsummer, the FCC issued an interim order, basically a six-month freeze on most existing terms for existing (not new entrant) CLECs. This provided Powell with enough time to consider his next career, the Bush administration with a reprieve on a tricky issue until after the election, and CLECs with a short extension of their businesses, while it preserved the regulatory uncertainty that had helped squelch new investment in CLECs.Independent

ISPs Face New Survival Threat from Bush reelection

After the Bush re-election in late 2004, the membership of the FCC was set for substantial turnover, but the agenda was clearly in place. In October 2004, the Commission granted the ILECs a few early holiday ponies. At BellSouth's request, it extended the TRO's ban on FTTH unbundling so that it also now applied to "fiber to the curb" (FTTC), so that copper loops no longer needed unbundling so long as they were attached to a fiber-optic-fed remote terminal whose copper distribution loops were all shorter than 500 feet. It had shortly earlier granted the same "unbundling relief" for copper loops within fiber-fed multiple dwelling units and mixed commercial-residential buildings. FTTC could *theoretically* provide adequate speed to provide near-fiber-like services, such as video on demand. But a real impact of the rule would be to make it far easier for ILECs to cut off CLEC access to unbundled loops. Verizon and SBC had been promising some degree of FTTH buildouts if they were not required to unbundle; BellSouth wanted, and got, the same benefits without ponying up the same cash. (It should come as no surprise that BellSouth had Powell's ear. His closest advisor, until her November, 2004, departure from the FCC 's

Office of Strategic Planning and Policy Analysis, was Jane Mago, wife of Robert Blau, BellSouth's Vice President for regulatory affairs.)

BellSouth then filed a petition for forbearance of the FCC's Computer II rules *and* their entire Title II common carrier obligation *per se* for all "broadband" access services. This was another approach to the same goal that Powell had proposed, but not acted on, three years earlier. A forbearance petition, unlike a rulemaking, takes effect automatically if the FCC doesn't dismiss it within a year. The BellSouth proposal would remove the ILEC's obligation to provide even the same raw bandwidth to ISPs that they provide to their own ISP offering (the Computer II obligation), and their obligation to make raw telecommunications service available to ISPs on request at just and reasonable rates (common carriage). With the erosion of loop unbundling taking away the CLEC's ability to service ISPs, the future prospects of independent ISPs became as seriously impaired as that of the CLECs. While dial-up remained available, even that came under more attack, as ILECs were succeeding in many cases in putting a prohibition on Virtual NXX operation into their interconnection agreements. Thus even low-speed dial-up ISP competition would only be available in major cities, not the bulk of the "red state" countryside. Would a tight ILEC/cable duopoly on ISP service encourage investment? With four more years of the Bush administration ahead, the outlook could not be said to be optimistic for anyone except the wealthiest incumbents.

Role Reversal: States Become Champions of Competition

When the Telecom Act first passed, proponents of competition supported greater federal power and less state authority, in large part because state regulators had often proven themselves, up to then, as opponents of competition. Although a few states had been supportive, others undertook rear-guard actions against the CLECs. This was an example of *regulatory capture*, when a regulatory body becomes largely beholden to the companies it is supposed to regulate. RBOCs and other ILECs had played that game well. It was the Iowa Utilities Board, after all, that was the lead plaintiff against the original unbundling rules.

But a funny thing happened as time went on. Local telephone competition did not simply save businesses money while resulting in higher rates for granny, as the incumbents had warned. Instead, UNE-P dramatically lowered rates for many consumers, even granny, providing ILECs with real price pressure for the first time. This proved to be politically popular, even in rural states. And indeed it was the rural states that were most dependent on UNE-P, because they were least likely to be able to support more elaborate facilities-based competition. So while the Powell FCC was contemplating the abolition of UNE-P in the TRO,

Kevin Martin's compromise, putting the onus on the states, was seen as very procompetitive. And after the D.C. Circuit vacated the TRO and potentially made most UNEs unavailable, numerous states took independent action to preserve them.

The CLEC sector had never been wealthy or strong. Early facilities-based CLECs had, for the most part, lost money. But that is typical for many pioneers; the first entrants in many businesses make the big mistakes and pave the path for more successful successors. With CLECs, though, the confluence of the boom-bust cycle with the regulatory changes that followed the 2000 election caused an even more perfect storm. A brief era of free-running capital backed by a friendly regulatory climate turned into an era of nearly nonexistent capital, feisty and litigious incumbents, and hostile federal regulators and courts. Many small, privately held CLECs eked out a living, changing course as necessary to avoid the storms, waiting for better days. But the optimism that had greeted the new sector just a few years earlier seemed almost like a distant dream, fading deeper into the past.

Endnotes

[1]   Maryland PSC order 71155; see also Katsouros, M., and C. Gump, "Telecommunications in the State of Maryland: Regulation and Competition," Office of Information Technology, University of Maryland, 1997, nts.umd.edu/~mark/630loca2.html.

[2]   Vogelsang, I., and G. Woroch, *Local Telephone Service: A Complex Dance of Technology, Regulation and Competition.* From *Industry Studies, 2nd Ed.*, 1998, Duetsch, L., and M. E. Sharp, eds., http://www.elsa.berkeley.edu/~woroch/dance.pdf.

[3]   This assumed the same carrier serving area design guidelines that ADSL was meant to support.

[4]   Sometimes rendered extended enhanced link or some other combination that leads to EEL; it is also formally referred to as a UNE combination. EELs were also created to support analog lines, with bundled channel banks; their prices, however, were very unattractive in most states.

[5]   There is no one universally accepted formal definition of virtual NXX, but it commonly refers to foreign exchange service in which the serving carrier has no physical customers in the remote exchange. ILECs usually provisioned FX service via the "long extension cord" method but on occasion did use a CLEC-like method in which a prefix code was rated to a physically distant exchange. This was done many years earlier, for instance, in the San Francisco area, where Pacific Bell put San Francisco-rated numbers in other exchanges that had a high demand for FX lines.

[6]   XCOM's architecture took advantage of IN2 capabilities to direct calls to the DMS-500 or directly to the RAS before local number portability would have made this easy, but this network capability was not made generally available to CLECs elsewhere.

[7]   Bell Atlantic Reply Comments in FCC docket 96-98, at 20. Also see hearing transcript on HR4445 before the House Subcommittee on Telecommunications, Trade and Consumer

Protection, June 22, 2000, http://www.nrri.ohio-state.edu/recipcomp/hearings%20on%20hr%204445.pdf.

[8]   The legal definition of "communications" is far more expansive than that of "telecommunications."

[9]   Massachusetts Department of Telecommunications and Energy docket 97-116-c.

[10]   Federal Communications Commission, "Local Telephone Competition: Status as of December 31, 2003," June 2000.

[11]   *United States Telecom Assoc. v. FCC,* 290 F. 3d 415, 426-27 (D.C. Cir. 2002), *cert. denied,* 123 S Ct. 1571 (2003).

[12]   *United States Telecom Assoc. v. FCC,* 359 F. 3d 554 (D.C. Cir 2004).

[13]   For example, *USTA II* noted that the *only* reason stated by the FCC for maintaining UNE-P was because of the high cost of hot cuts. The record was quite a bit more expansive. Also see Goldstein, F., and J. Marashlian, "USTA v. FCC: A Decision Ripe for the Supremes," ISP Planet, March 25, 2004, isp-planet.com/perspectives/2004/usta_v_fcc.html.

[14]   Precisely what the ILECs offered cannot be stated, because their offers, while made to thousands of CLECs, were technically under nondisclosure.

10

Focusing on the Bottom Line

The great telecom meltdown did not occur because of any one new law or rule, any one new technology, or any one company's actions. It was the result of a confluence of events, a set of small storms that combined into one very big one. But just as large hurricanes do not usually form unless the air and sea are at the right temperature, large financial storms require certain environmental conditions to be present. Businesses respond to many inputs, but most are ultimately responsible to their investors. When investors set aside common sense and instead look for reasons to justify irrational behavior, they are creating the climate for stormy weather. Sometimes a major storm may begin by looking like a good sailor's breeze, but those who pay attention to the signs can usually get out of harm's way on time.

The meltdown provides many fine examples of bad ways to use one's money. But the underlying truths are fairly simple. Technology, business, and regulation move, more or less, together. Technological progress usually leads. Regulators need to respond appropriately or risk having their regulatory schemes collapse under the weight of arbitrage or unforeseen competitive forces. The details of the economy change, but the underlying principles do not. There never is, and never will be, a real "new economy." And thus simple laws of business still apply: Companies need to eventually make a profit or they cannot continue to operate. They must focus on the bottom line, not whatever convenient metric happens to make them look good. Engineers and scientists, but not accountants, are the ones who should be creative.

Asset Valuation is Risky

At the heart of the meltdown was the vast level of investment in telecommunications facilities made during the 1990s. The decade began with a recession, with little new investment and little new motivation for investment in technology-based companies. The minicomputer industry had crashed just a few years earlier, and commoditization was driving down margins across the computer industry. Corporations were building private data networks, but most telecom-sector growth was fairly slow and steady, with mobile telephones just starting to penetrate the mass market. It was not until the Internet went public that the impetus existed for large-scale new expenditures in both telecommunications plants and in facilities to make use of it.

A key problem with many of these investments was that the assets were simply not worth what they cost to build, in large part because a glut of construction drove down prices and spread the business too thin. Of course no one builds such assets knowingly! But asset valuation, always difficult, was often being driven to distraction by stock market bubbles, inflated demand forecasts, ignorance of other new competitors, and valuation formulas that were more likely designed to make investors open their pocketbooks than to deliver a true answer.

Assets can be valued in different ways, of course. One approach is to look at book value: Take the amount invested and subtract depreciation. This works for tax purposes, but it makes no attempt to balance supply with demand. Its main benefit is simplicity. Investors do not, however, shell out cash because they will just end up with book value. Another metric is enterprise value: Look at a business' revenues and profits, apply some time-value-of-money metrics, and determine a net present value. That is usually much closer to what a business is really worth. But it has much more room for error, because it depends on predictions of the future. During the boom years, entrepreneurs often got help from industry analysts who willingly exaggerated future demand. This provided higher forecasts for future profit, giving their plans a higher enterprise value.

And at the end of these multiyear profit forecasts, enterprise valuation formulas include a *terminal value*, typically a multiple of forecasted revenues, based on the assumption that a stable, mature company's value can be approximated that way. So if a business was forecasting 10 years of high top-line growth, its terminal value could be extremely high, and thus the net present value (that is, with the estimated future value discounted for the time value of money) could still be dominated by that top-line forecast. Valuations depending only on realistic future *profit* would have been lower, of course, but they were not fashionable during the boom years.

Many investors did not even think about valuation in any kind of normal sense. During the dotcom boom, the most flagrantly overvalued companies, such as Web site operators, were frequently judged by made-up or secondary metrics such as "eyeballs" or "page hits." Those may have been interesting metrics' for the operators' egos, but they did not necessarily translate to a bottom line. Within the telecommunications industry, companies often hyped their investments by using metrics that, while impressive, were fundamentally irrelevant. CLECs and DLECs, for instance, sometimes talked about accessible lines—the number of telephone lines in central offices where they had customers. But the fifth DLEC in a central office with 50,000 residential and 10,000 business subscribers was probably going to get less business than the first one in a CO with 5,000 of each. And those supposedly accessible line counts did not always take into account the Achilles heel of DSL—loop qualification— which often rendered more than half of the lines in some COs functionally inaccessible. CAPs similarly talked about miles of plant, or the number of businesses along their routes, without taking competition into account. Venture capitalists who thrived on such measurements were doing almost everyone a disservice, though the ones who managed to sell shares to the public on time tended to do pretty well for themselves.

Accounting Was Scandalous

The boom years were also the years of some of the biggest accounting scandals in history. And the company that personified the telecom sector's rapid growth, WorldCom, was one of the most flagrant practitioners of creative accounting. Other major scandal-ridden companies were also involved in telecom. Enron became the poster child for accounting fraud, taking down giant audit firm Arthur Andersen with it. Although not primarily a telecom shop, its Enron Broadband unit had promised big profits based on trading bandwidth, just as the parent company had traded energy. Its claims were, in large part, lies, designed to support an excessive stock price. Adelphia Communications' and Tyco International's scandals were less about fictitious accounting than about CEOs who skimmed a little off the top, but such behavior would not have been possible had proper fiscal controls been in place. And those are just companies that got caught crossing the legal boundary. Even companies that did not violate any rules often depended on questionable accounting practices such as emphasizing "pro forma" rather than generally accepted accounting principles (GAAP) profit numbers. The usual definition of "pro forma" was, well, whatever the reporting company felt like, since it was not a term with a legally specific meaning.

Investors and Entrepreneurs Played Each Other for a Loss

The telecom boom was largely about new business start-ups, some small, but some quite large. Start-ups generally happen when an entrepreneur gets capital from investors. The first investors, the "angels," provide seed money to allow the entrepreneur to flesh out the idea, while the first major funding typically comes from venture capitalists. (To be sure, many companies were started and funded by existing companies or by serial entrepreneurs who provided their own angel or even venture funding.) The venture capital process has some fairly standard practices, often involving multiple rounds, and typically ends when the venture firm either pulls the plug or cashes out. The exit strategy for a successful firm would typically be a public stock offering or a sale of the company. If the venture firm pulled the plug by refusing further funding, the entrepreneur might attempt to find additional funding elsewhere, but if that failed, bankruptcy usually resulted. During the boom, many entrepreneurs committed the fatal error of anticipating further rounds of funding that never materialized, and spending down their cash in order to grow faster, without having enough to last until the cash flow went positive. They were not prepared for the boom to end. But booms always end. That is what the business cycle is all about.

Once an early investor funds a start-up, it has an interest in making the company succeed in the short run, enabling it to sell its stake to other investors at a profit. Investment bankers likewise have an interest in stirring up the maximum interest for companies they are peddling. So they want a good "story." And successful entrepreneurs were often very good storytellers. But in complex industries like telecommunications, which are sensitive to both fast-moving technology and fast-changing regulation, it can be hard for even a professional investor to tell a good story from a cock-and-bull story.

At the biggest financial firms, that role fell to the analysts. But while many worked hard for their once-sumptuous pay, some of them turned out to be rather honesty-challenged. This was not always their own fault: If a firm's stock brokerage arm employed an analyst to help its customers determine what stocks to buy, and its investment banking arm wanted business from companies whose stocks were being covered by the analysts, then the analysts were under pressure to please the companies, not their own customers. Best-known of these analysts in the telecom sector was Salomon Smith Barney's Jack Grubman, who famously kept his buy rating on WorldCom after the company's stock was moving into free fall—and whose daughter got into a very selective preschool after he gave a suspiciously favorable report on AT&T. Salomon's parent company Citigroup ended up paying $2.65 billion to investors who invested in World-Com while Grubman was recommending it [1]. Some story indeed!

WorldCom and the Limits to Mergers

The accounting fraud that finally did in WorldCom was in large part tied to the complex regulatory system that applied to the domestic telephone industry. It had been booking access charges—the subsidy-laden payments it made to local exchange carriers for the use of their networks—as capital investments. This would have put a burden on future earnings, to be sure, but it enabled the company to maintain its reported margins in the short term. Treating access charges as capital was obviously improper, albeit reminiscent of the way New York City's municipal government had, at one point in the 1970s, capitalized some of its entitlement payments as "human capital." (That let the city borrow to cover the year's expenses, but it left them deeply in debt, requiring a subsequent bailout.) WorldCom's sins ended up at the same font of absolution as so many others', the bankruptcy court, and, indeed, in criminal court as well.

But before WorldCom got there, it faced a more serious problem that the financial community was largely remiss in addressing. It had grown by acquisition, and continued to grow rather quietly until, after the Wiltel acquisition in 1995, it got to be the fourth-largest interexchange carrier. Its high growth rate helped maintain the stock multiple, which helped it acquire more companies. It succeeded in buying MCI, vaulting itself to number two, but then had nowhere left to go. Even if it *had* gotten hold of Sprint, which was by that point rather smaller than the combined MCI WorldCom, and even if it *had* really been making the kind of profits that it was reporting, Bernie Ebbers' company would have had to shift gears to concentrate on improving operations and margins. And that requires a very different kind of manager, and a different kind of corporate culture. That was not the type of corporate culture that the boom years, with their emphasis on growth rather than profitability, cultivated.

AT&T Acted on Faith in WorldCom's Numbers

WorldCom turned in impressive, if falsified, results even after long distance prices were falling and their competitors' margins were getting squeezed badly. This was not only a problem for WorldCom's many deceived investors but for its competitors as well, because other companies made their own decisions based on a benchmark of WorldCom's reported profits. AT&T reportedly divested itself of its wireless and cable operations—its two best potential growth areas, in an era of declining long distance revenues—because WorldCom was reporting a better return on its assets.

AT&T itself had been undergoing change at a rate that the stodgy old corporation was having a hard time dealing with. Its bloated basic structure, its DNA as it were, dated back to its days as a huge monopoly. With the RBOCs divested, the remaining company had to learn to live in a competitive world. That was hard enough, though it was a transition that many companies have had to deal with. But with the Telecom Act, the main customers of its manufacturing arm were suddenly its competitors. For that and other reasons, it spun off Lucent Technologies, and with it most of Bell Labs. That too was traumatic. CEO Michael Armstrong made a last gasp for glory by buying Tele-

Communications Inc., turning overnight into the biggest cable company. And of course it paid far too much—Armstrong was no master of valuation. But it could have continued with its strategy of using the cable plant for triple play, making itself relevant again as well as a worthy competitor to the RBOCs. But instead, it underwent a reorganization that led to the spin-off of both its cable and wireless units, leaving old Mother Bell just a ghost of its glorious former self. Armstrong himself was gone, replaced by David Dornan, who had not come up through the ranks of management; he would have to find relevance for the smaller company or be forced to find an end game.

Global Double Crossing

Global Crossing was another boom-era company whose business was in a natural meltdown, as its heavy investments in undersea cable were not paying off. Like Enron, it retained Arthur Andersen as its audit firm. Like Enron, it ended up in bankruptcy, after its founder, Gary Winnick, had made $700 million for himself by selling company stock into the bull market. (Hence this raised a familiar question—what did he know and when did he know it?) The company was also lavish with politicians, including former presidents Bill Clinton (a gift to his presidential library) and George H.W. Bush (stock in lieu of a speaking fee) [2]. The company's fall had many causes, but it apparently used questionable accounting of its wholesale sales to other telecommunications companies. It made "even trades" of IRUs with other carriers, booking the sales as revenue. This provided its trading partners with their own opportunity to book the trades as sales. It talked about IRU sales as providing "cash revenue," an impressive-sounding term not found in standard accounting practices but which exaggerated apparent earnings [3]. It apparently swapped $150 million on paper with 360Networks, with no cash actually changing hands [4].

And indeed others did the same thing. Qwest had to restate several years' past earnings after its IRU deals were found to have been accounted for improperly. (Speaking of trading partners, Global Crossing did, after all, have IRUs on roughly half of Qwest's backbone fiber.) The company's former US West ILEC unit was subject to relatively strict state regulation, and it was a cash cow for the rest of the company, keeping it out of bankruptcy, while the long distance unit had tried to exaggerate sales and later hide losses.

New Services Need to Fit Into a Food Chain

Another problem that hurt many companies, both equipment suppliers and service providers, was a failure the *food chain* that their products needed to fit into. A food chain, in nature, describes how the smallest forms of life are eaten by successively larger ones. In a technology business, it refers to a set of dependencies that companies, technologies, and products have on others. If a food chain is broken in nature, higher life forms will have to adapt or they will not survive. If a food chain is broken in business, its effects can be felt both above and below (Figure 10.1).

The bottom of the tech-business food chain is raw *technology*. Fiber optics, radio communications, digital switching, and Internet protocols all fall into this area. It is sometimes sexy, and it is certainly necessary, but it is not profitable unless the rest of the chain above it falls into place. Developers of technology that fail to assess the likelihood of a complete food chain above them are thus taking a great risk. Technology is then used to create individual *products* that are in turn parts of more complex *systems*. If the systems do not need the technology or products, the technology will have trouble selling.

Systems are used to create *infrastructure* and, in turn, *services*. A service provider thus needs to have a viable chain of systems, products, and technology to support its plans. But it also needs to support *applications* that fill *market needs*. Services and applications that do not fill market needs as well as competitors will have trouble competing. And almost every layer is sensitive to price. Greater product differentiation can sometimes allow a product to sell at above-commodity prices, but it can also make it harder to fit into real market needs. It is a tough balancing act but a necessary one.

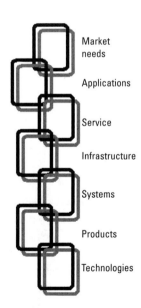

Market needs

Applications

Service

Infrastructure

Systems

Products

Technologies

Figure 10.1 Food chain for technological products. A successful product requires all links in the chain to be complete.

Competitive Realities Will Change

Here is the key to the whole meltdown: Competitive industries do not work the same way as monopolies. It is simply not easy for an industry to make that shift. It is often, however, necessary, because monopolies are the exception, not the rule, in any vibrant economy. And as nature abhors a vacuum, technology abhors a monopoly and tries to work around it.

Monopolies can often get a lot more out of an investment than can players in competitive markets. Monopoly telephone companies installed central office switches that lasted 40 to 50 years. Monopolies depreciated the cost of computer memory chips over decades. Monopolies worried about service life. Competitive companies deal with economic life, not service life. Competitors know that plant or a product is retired when it is no longer competitive, even if it still works like new. Competitors know that no matter how big the investment, there is no guarantee of profit or even significant revenue. Given those contrasts, experienced telecom industry managers and investors both had trouble coping with the transition.

Competition is usually good. It is usually good for consumers, and it is usually good for suppliers. It can even be good for a market participant: Do Toyota Motors or General Electric complain about not having a monopoly? They pick markets and focus on them, while competitors keep them on their toes and give them new ideas. Competitive companies do not expect to be all things to all people, to supply all parts of a broadly defined marketplace; in contrast, ILECs grew up with that mission. The many changes in the telecommunications industry's regulatory regime were driven by technology and had to be met by appropriately revised business thinking. Inappropriately revised business methodologies caused absurdities of overinvestment and led to the meltdown. Companies that sat tight on their dollars and stuck to conservative financial practices generally survived the meltdown; many prospered. There is no shame in being the survivor that gets the resulting bargains.

Old Dinosaurs Die Hard

Much of the telecom boom investment was spent on companies whose goal was to displace well-established competitors, be they equipment manufacturers, long distance companies or local exchange carriers. Entrepreneurs and their investors often evaluated these companies' products and operations and decided that they were vulnerable, often because they were behind the curve on technology. What they failed to count on was the market inertia supporting existing vendors. Customers had existing relationships and were often slow to risk new suppliers. Ma Bell's PBX business led the market for some years after Carterfone, even though its products were often inferior and costlier; it took the Computer II decision, forbidding the local exchange carriers from distributing such equipment, to end that dominance. Millions of consumers stuck with AT&T's long distance service after divestiture simply because they were not interested in choosing a replacement; they did not even switch to a less-costly AT&T plan. AT&T benefited by this inertia. And while the coming of CLECs was greeted enthusiastically by perhaps a quarter of the total market, which was looking for any credible alternative to the old monopolies, many more subscribers simply did not care enough to switch unless the offer was truly compelling. And that included factoring in the risk of dealing with a new, not-yet-well-established vendor.

On occasions that the incumbents did feel serious market heat, they had many resources at their disposal. Sometimes they were able to forestall losing sales to competitors by going straight to senior executive levels; a CEO is often more amenable to pressure from an incumbent vendor than an information technology manager or even a CIO. In some cases, especially sales to large businesses, they often offered selective discounts. Perks like Super Bowl tickets to big customers did not hurt either. In some cases, especially the ILECs, they simply went to the courts or regulators to have the rules changed. Big, established companies are usually more able to do that than start-ups. They may have been dinosaurs, but the real dinosaurs must have had something going for them; they lasted for more than a hundred million years.

The Next Big Thing Usually Is Not

Technology investors are often hunting for the "next big thing." This occasionally results in a big payoff, especially for the investor who gets in early and gets out near the top. But of course it is hard to know when the market is topping off until it is too late. Lots of next big things turned out not to be. In the late 1970s and early 1980s, it was integrated voice-data PBXs. Then for a brief moment, it was supposed to be integrated voice-data workstations. Then in at least some parts of the world it was sure to be ISDN, then ATM. Then it was the Web, and any business beginning with the letter "e." And anything having to do with fiber optics.

Venture capital is a high-stakes game. A venture firm succeeds if a small percentage of its investments pan out big; venture professionals are also experts at jumping off a moving train. Most small investors cannot do this. They become wedded to a stock, or they hang on too long, hoping its price will recover. Individual investors were some of the biggest victims of the meltdown. They often received false signals from the press, analysts, and even friends and fellow investors. Press reports do not always turn into profits. Reporters are looking for a good story of a different kind. Sometimes they are focusing on a technology that is beneficial to its users but not necessarily its investors. Usually they have only a cursory knowledge of a topic and lack the in-depth understanding that would make future failures more evident. And when something gets too much publicity, too many people want to buy in, forcing up the price of equities beyond sustainable levels. Fads are thus the worst investments; the price cannot help but fall once the excitement moves elsewhere.

The great telecom meltdown was the result of many things. Most of all, it resulted from the madness of crowds. An industry that was facing great change provided many investment opportunities, both good and bad. The boom years of the late 1990s were a time of free-flowing capital, and the telecom industry was ready to absorb it. The problem was not competition, or reregulation, or too much technology. It was emotion, the inability to set aside excitement and think rationally when making investment decisions. It happened to tulips, it happened to telecom, and it will happen again, somewhere else. It always does.

Endnotes

[1]   "Citigroup Agrees to a Settlement Over WorldCom," *New York Times*, May 11, 2004.

[2]   Ackman, D., "House Committees to Investigate Global Crossing," Forbes.com, March 13, 2002, forbes.com/2002/03/13/0313topnews.html.

[3]   Long, J., "Tricks of the Trade: Accounting for Swaps," , *Phone Plus*, September 2002, http://www.phoneplusmag.com/articles/291carrier2.html.

[4]   Long, J., "Tricks of the Trade: Accounting for Swaps," , *Phone Plus*, September 2002, http://www.phoneplusmag.com/articles/291carrier2.html.

List of Acronyms

1G	first-generation
2G	second-generation
3G	third-generation
3GPP	3d Generation Partnership Project
ABI	American Bell Inc.
ACD	automatic call distribution
ADL	Arthur D. Little Inc.
ADM	Add-on Data Module
ADSL	asymmetric digital subscriber line
AMI/ZCS	alternate mark inversion, zero code suppression
AMPS	Advanced Mobile Telephone Service
ANS	Advanced Networks and Solutions
ANSI	American National Standards Institute
ATC	Advanced Telecom Corp.
ATM	asynchronous transfer mode
AUP	acceptable use policy
B8ZS	bipolar 8-zero substitution

BBS	Bulletin Board System
BETRS	basic exchange telecommunications radio service
B-ISDN	Broadband ISDN
BOCs	Bell Operating Companies
BRI	basic rate interface
BSD	Berkeley Software Distributions
BT	British Telecom
BTA	basic trading area
CAP	competitive access provider
CAPAC	cable packet controller
CAPI	Common ISDN Application Protocol Interface
CATV	cable television
CCITT	International Telephone and Telegraph Consultative Committee
CDMA	code division multiple access
CDPD	cellular digital packet data
CIX	Commercial Internet Exchange
CLEC	competitive local exchange carrier
CLNP	Connectionless Network Protocol
CMTS	cable modem termination system
CO	Central Office
CPE	customer-premise equipment
CPU	Central Processing Unit
CSA	carrier serving area
CSMA/CD	Carrier Sense Multiple Access with Collision Detection
CWDM	coarse wavelength division multiplexing
DAA	data access arrangement
DEC	Digital Equipment Corp.

DLEC	data-oriented CLEC
DMI	Digital Multiplexed Interface
DNS	Domain Name System
DOCSIS	Data Over Cable System Interface Specification
DoD	Department of Defense
DS-1	digital signal level 1
DSL	digital subscriber line
DSLAM	DSL access multiplexor
DWDM	dense wavelength division multiplexing
EBITDA	Earnings before Income Taxes, Depreciation, and Amortication
EDGE	Enhanced Data Rates for GSM Evolution
ELEC	Ethernet Local exchange carrier
ENFIA	Exchange Network Facilities for Interstate Access
ESP	Enhanced Service Provider
ESS	Electronic Switching System
EUCL	end user common line charge
FCC	Federal Communications Commission
FDDI	fiber distributed data interface
FJ	Final Judgment
FSS	fully separate subsidiary
FTTH	fiber to the home
FX	foreign exchange
GAAP	generally accepted accounting principles
GOSIP	Government OSI Profile
GPRS	general packet radio service
GSM	Global System for Mobile
HDSL	high-bit-rate digital subscriber line

HDTV	high-definition television
HFC	hybrid fiber-coax
HFE	high frequency element
HP	Hewlett-Packard Inc.
HPFL	high frequency portion of the loop
HSCSD	high-speed circuit-switched data
IDSL	ISDN digital subscriber line
ILEC	Incumbent local exchange carrier
IMP	intermachine processors
IMT-2000	International Mobile Telecommunications-2000
IMT	intermachine trunk
IMTS	improved MTS
IN	Intelligent Networks
IOF	interoffice facility
ION	integrated optical network
IP	Internet Protocol
IPRS	Internet Protocol Routing Service
IRU	indefasible right to use
ISDN	integrated services digital networks
ISO	International Organization for Standardization
ISP	Internet service providers
ITU-T	International Telecommunication Union
IVD	integrated voice and data
IXC	interexchange carrier
KSU	key service units
LAN	local area network
LATA	local access and transport area

LEC	local exchange carriers
LMDS	Local Multipart Distribution Service
LNP	local number portability
MAC	media access control
MAE-East	Metropolitan Area Ethernet East
MAE-West	Metropolitan Area Ethernet West
MCI	Microwave Communications, Inc. originally
MFJ	Modified Final Judgment
MFN	Metromedia Fiber Network
MFS	Metropolitan Fiber Systems
MPEG	Motion Picture Experts Group
MPL	multischedule private line
MSA	metropolitan statistical area
MSO	multiple system operators
MTA	major trading area
MTS	message toll service
MTS	mobile telephone system
MTSO	mobile telephone serving offices
NCP	network control protocol
NFSnet	National Science Foundation network
NREN	National Research and Education Network
NTS	non-traffic-sensitive
OC	optical carrier
OEM	original equipment manufacturer
OSI	Open Systems Interconnection
OSIRM	OSI Reference Model
OSS	operational support systems

PARC	Palo Alto Research Center (Xerox)
PBX	Private Branch Exchange
PCS	personal communications service
PDA	personal data assistant
PIC	primary interexchange carrier
PoP	point of presence
PRI	primary rate interface
PTT	Post, Telephone, and Telegraph agencies
PUC	Public Utilities Commission
QoS	quality of service
RAS	Remote Acess Server
RBOCs	Regional Bell Operating Companies
RCC	radio common carrier
RSA	rural service area
SBS	Satellite Business Systems
SCC	specialized common carrier
SDH	synchronous digital hierarchy
SDSL	symmetric digital subscriber line
SMSA	standard metropolitan statistical area, later renamed MSA
SNA	Systems Network Architecture
SONET	Synchronous Optical Network
SPCC	Southern Pacific Communications Company
SPID	service profile identifier
STM	Synchronous Transport Module
STS	Synchronous Transport Signal Level
TCG	Teleport Communications Group
TCI	Tele-Communications Inc.

TCP/IP	Transmission Control Protocol/Internet Protocol
TDM	time division multiplexing
TDMA	time division multiple access
TELRIC	Total Element Long Run Incremental Cost
TIRKS	Trunk Integrated Record Keeping System
TRO	Triennial Review Order
UMTS	universal mobile telephony system
UNE	unbundled network elements
UNE-P	UNE Platform
USTA	United States Telecom Association
UUCP	Unix to Unix Copy Protocol
VDSL	very-high-speed digital subscriber line
VLSI	very large scale integration
VOD	video on demand
VoIP	Voice over Internet Protocol
VSAT	very small aperture terminal
VTPP	Variable Term Payment Plan
WAP	Wireless Access Protocol
WATS	wide area telephone service
W-CDMA	wideband CDMA
WDM	wavelength-division multiplexing
WLL	wireless local loop

About the Author

Fred R. Goldstein is principal of Ionary Consulting, where he advises companies on technical, regulatory, and business issues related to the telecommunications and Internet industries, especially in areas where the two overlap. He has helped numerous competitive carriers navigate the start-up process, helping them deal simultaneously with technical, regulatory, and business issues. He assists service providers in network design, business modeling, planning, and technical architecture. He has frequently been an expert witness in intercarrier compensation and network interconnection cases. He has worked with enterprise networks on a wide range of matters such as backbone network design, voice systems planning, and traffic engineering.

Prior to starting Ionary, Mr. Goldstein worked for Arthur D. Little Inc., BBN Corp., and Digital Equipment Corp. A graduate of Skidmore College, he is the author of numerous articles and the book *ISDN in Perspective*, has served on standards committees in areas such as ATM networks and Frame Relay, and has taught courses for Northeastern University and National Technological University.

Index

Recent Titles in the Artech House Telecommunications Library

Vinton G. Cerf, Senior Series Editor

Videoconferencing and Videotelephony: Technology and Standards, Second Edition, Richard Schaphorst

Visual Telephony, Edward A. Daly and Kathleen J. Hansell

Wide-Area Data Network Performance Engineering, Robert G. Cole and Ravi Ramaswamy

Winning Telco Customers Using Marketing Databases, Rob Mattison

WLANs and WPANs towards 4G Wireless, Ramjee Prasad and Luis Muñoz

World-Class Telecommunications Service Development, Ellen P. Ward

For further information on these and other Artech House titles, including previously considered out-of-print books now available through our In-Print-Forever® (IPF®) program, contact:

Artech House
685 Canton Street
Norwood, MA 02062
Phone: 781-769-9750
Fax: 781-769-6334
e-mail: artech@artechhouse.com

Artech House
46 Gillingham Street
London SW1V 1AH UK
Phone: +44 (0)20 7596-8750
Fax: +44 (0)20 7630-0166
e-mail: artech-uk@artechhouse.com

Find us on the World Wide Web at:
www.artechhouse.com